Werner Rudolf Cramer

Interference Pigments

AF818583

Also of Interest

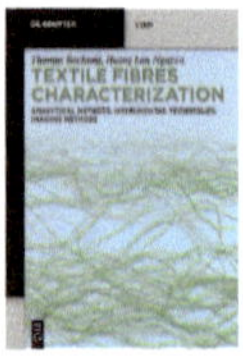

Textile Fibres Characterization
Analytical Methods, Instrumental Techniques, Imaging Methods
Thomas Bechtold, Huong Lan Nguyen, 2026
ISBN 978-3-11-160950-8, e-ISBN (PDF) 978-3-11-160966-9

Inorganic Pigments
Gerhard Pfaff, 2023
ISBN 978-3-11-074391-3, e-ISBN (PDF) 978-3-11-074392-0

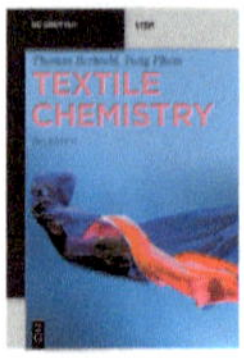

Textile Chemistry
Thomas Bechtold, Tung Pham, 2023
ISBN 978-3-11-079569-1, e-ISBN (PDF) 978-3-11-079573-8

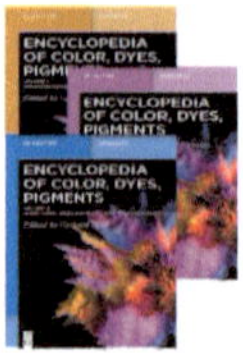

Encyclopedia of Color, Dyes, Pigments
Edited by Gerhard Pfaff, 2022
Volume 1: Antraquinonoid Pigments - Color Fundamentals
ISBN 978-3-11-058588-9, e-ISBN (PDF) 978-3-11-058807-1
Volume 2: Color Measurement - Metal Effect Pigments
ISBN 978-3-11-058684-8, e-ISBN (PDF) 978-3-11-058710-4
Volume 3: Mixed Metal Oxide Pigments - Zinc Sulfide Pigments
ISBN 978-3-11-058686-2, e-ISBN (PDF) 978-3-11-058712-8
Also available as Set: ISBN 978-3-11-058881-1

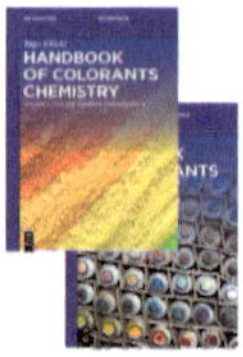

Handbook of Colorants Chemistry
Ingo Klöckl, 2023
Volume 1: Dyes and Pigments Fundamentals
ISBN 978-3-11-077699-7, e-ISBN (PDF) 978-3-11-077711-6
Volume 2: in Painting, Art and Inks
ISBN 978-3-11-077700-0, e-ISBN (PDF) 978-3-11-077712-3
Also available as Set: ISBN 978-3-11-114610-2

Polymer Surface Characterization
Edited by Luigia Sabbatini, Elvira De Giglio, 2022
ISBN 978-3-11-070104-3, e-ISBN (PDF) 978-3-11-070109-8

Werner Rudolf Cramer

Interference Pigments

Color and Measurement

DE GRUYTER

Author
Werner Rudolf Cramer
Hafenweg 22
48155 Münster
Germany
info@wrcramer.de

Cover image - Car body with raindrops
The color of interference pigments also depends on the refractive index of the environment. This interference pigment shows orange and red on a car body. The raindrops have a higher refractive index than air, which is why the color shifts to yellow and green.

ISBN 978-3-11-143842-9
ISBN 978-3-11-143920-4 (PDF)
ISBN 978-3-11-143946-4 (E-PUB)
DOI https://doi.org/10.1515/9783111439204

Library of Congress Control Number: 2026937901

Bibliographic information published by the Deutsche Nationalbibliothek
The Deutsche Nationalbibliothek lists this publication in the Deutsche Nationalbibliografie; detailed bibliographic data are available on the Internet at https://dnb.dnb.de.

© 2026 Walter de Gruyter GmbH, Genthiner Straße 13, 10785 Berlin, Germany

All rights reserved. No part of this publication may be reproduced, translated, stored in a retrieval system, or transmitted in any form or by any means, electronic, mechanical, photocopying, recording or otherwise, without prior written permission from the publisher. Text and datamining for commercial purposes requires the publisher's permission.

De Gruyter and Walter de Gruyter GmbH are part of De Gruyter Brill.
www.degruyterbrill.com

Questions about General Product Safety Regulation: productsafety@degruyterbrill.com

Cover illustration: Werner Rudolf Cramer
Printing: CPI books GmbH, Leck

FSC www.fsc.org MIX Papier | Fördert gute Waldnutzung FSC® C083411

Contents

Foreword

My first encounter with interference pigments date back over 50 years. At that time, I saw for the first time the engine covers of a vehicle that were painted with pastel colored mixtures of interference pigments and colored pigments. I laid these hoods out like UFOs on a wet, black-brown field and photographed them (Figure 1).

Figure 1: Beetle hoods with interference pigments.

I then used the first car paint with a red interference pigment to paint two wooden balls (10 cm in diameter), which I still have today and enjoy showing off (Figure 2). The next big step was repainting three vehicles with interference pigments. To do this, my vehicles were completely dismantled and first painted with white primer, then with white interference pigment, and finally with clear coat. I applied adhesive tape to this clear coat to paint the corresponding shapes. I mixed white interference pigment with various colored pigments to use as basecoats. I mixed white filler with the same colored pigments as primers. I then sealed all the colored stripes with up to eight layers of clear coat, sanding between layers to achieve a uniformly smooth surface. This made the colored stripes look as if they were embedded in glass (Figure 3).

I successfully exhibited these three cars at numerous exhibitions. During this creative phase, I designed further vehicles with interesting designs using interference pigments. Despite my artistic success, I wanted to learn more about interference pigments. That is why I turned my attention to the scientific and technical aspects. Working with a manufacturer of these pigments led to an intensive study of these pigments and their optical behavior. At the same time, I was invited by a manufacturer of optical devices to collaborate on the development of a new multi-angle measuring device specifically for

https://doi.org/10.1515/9783111439204-203

Figure 2: Pearl red on a green substrate.

Figure 3: Design coatings with interference pigments.

interference pigments. With this laboratory device, the illumination and observation angles could be adjusted independently of each other in 5° steps. Working with both manufacturers gave me an enormous boost in knowledge. My knowledge of interference pigments and their properties have multiplied as a result.

At the end of the 1990s, I got together with American colleagues in Santa Rosa. This marked the birth of the ASTM Committee E12.12 on "Gonioapparent colors," which defined the standard method for measuring interference pigments in E2539 and published it for the first time in 2008. I was then also involved in the development of a new portable measuring device based on E2539. This laboratory device opened almost endless possibilities for me to try out measurement geometries, and thus learn more about interference pigments and their properties. Later, I realized that I was one of the few who could work with the many measurement geometries of the laboratory device, but no one could verify my results. So, I found ways to reproduce the results with the portable measuring devices.

I have enjoyed sharing my knowledge in numerous publications worldwide and in many lectures and presentations. This includes this book, in which I have summarized a wealth of interesting information about interference pigments and their optical properties. I hope that my enthusiasm for interference pigments is reflected in the book.

1 Introduction

People have always been fascinated by colors and effects. There are many examples in nature where the optical property of interference plays a major role. Butterfly wings, beetle shells, and mother of pearl are interesting examples in nature, as are soap bubbles and oil films on water.

Small pieces of mother of pearl were used in Japanese lacquer art, and fish silver was obtained from fish scales for many decorative applications. In 1934, a patent based on crystalline lead thiosulfate was filed as a substitute for fish silver.

At the beginning of the last century, attempts were made to produce similar effects artificially. These attempts led to bismuth oxychloride and iodide, which when manufactured using the appropriate process, became flake-shaped platelets. A large proportion of these pigments are still used in the cosmetics industry today. They are not suitable for automotive paints due to their low light and weather stability. They are also used in the production of Majorcan pearls.

There have also been experiments with basic lead carbonate, whose thin crystal flakes reveal interference colors.

In 1969, DuPont signed a contract with the German company Merck and the American company Mearl. Through cross-licensing, both companies acquired the manufacturing rights for interference pigments based on mica flakes. Merck has since sold its interference pigment business to the Chinese manufacturer Chesir (Global New Material International (GNMI)), while Mearl was first taken over by Engelhardt, then by BASF Corporation, and finally sold to the Japanese manufacturer Sun Chemical. Historical color charts from DuPont and Merck show the original interference pigments (Figure 4, Figure 5).

In a wet chemical process, mica flakes are coated with titanium hydroxide and sintered at 900 °C, whereby the titanium hydroxide is converted into titanium dioxide. Depending on the process duration, a thin layer thickness results in white interference pigments, while a longer process duration and a thicker layer thickness result in yellow, red, blue, and finally green interference pigments (Figure 6).

If iron oxide is used as the coating material, red interference pigments are obtained. Combination pigments with titanium and iron oxide coatings produce yellow-gold pigments, while chromium oxide produces green pigments. In addition to synthetic mica, interference pigments with silicon dioxide (silica), borosilicate, and aluminum oxide (alumina) as carrier materials are also available today.

In addition to wet chemical manufacturing processes, interference pigments are also produced in a high vacuum. They contain an aluminum core on which highly refractive layers are built up.

The descriptions and illustrations in this book are based on approximately 16,000 specially produced samples, most of which were created as spray outs, but some as drawdowns.

https://doi.org/10.1515/9783111439204-001

Figure 4: Historical color charts from DuPont.

These samples were measured colorimetrically using various measuring devices and evaluated. The results are mainly presented as reflection curves. They show the physical and optical properties of the interference pigments and their mixtures. The L*a*b* color values reflect our color perception, and thus indicate the color component of the interference pigments.

The success of interference pigments, particularly in automotive painting, was accompanied by a change in painting processes in OEM and repair painting. Whereas effect pigments were originally applied in a 2K process as single-coat systems, from the 1970s onwards they were applied in a two-coat process. In this process, a base coat containing the pigments is first sprayed on and allowed to dry physically. This is then sealed with clear coat to protect against mechanical and chemical influences. This process is safer and less complicated than the 2K process, especially for repair coatings. The first

Muster auf dunklem u. hellem Untergrund	Farbe	ungefähre Schichtdicke in mµ (berechnet)	Reaktions-dauer in Std.	Vers. Nr. Ti W
	silberweiß	40	8.5	353
	hellgoldgelb	55	11.5	353
	dunkelgoldgelb	70	14.0	353
	orange	80	15.5	287
	rot	90	16.5	288
	purpur	100	18.5	291
	violett	110	19.0	353
	blau	120	20.0	289
	blaugrün	140	21.5	286
	grün	150	23.0	285

Figure 5: Historical color chart from Merck.

Figure 6: First color series with mica pigments from white to green.

vehicles with white interference pigments were offered in Europe by Renault and Volvo in the mid-1980s. This was followed by a wave of blends of interference pigments with colored pigments. From around the mid-1990s, three-way mixtures of interference, colored, and aluminum pigments were used. These compositions still dominate the world of car colors today.

Portable measuring devices that measure the panel or bodywork on the assembly line from multiple angles also contributed to this success. This enabled colors to be defined more precisely and incorporated into the car manufacturers' specifications.

This interplay of interference pigments, application methods and materials, and color measurements is now commonplace and state of the art. It should not be forgotten that interference pigments have also found their way into other areas such as plastics and cosmetics. They have developed their fascination there, too.

The optical properties of interference pigments are illustrated using their reflection curves and L*a*b* charts. A separate chapter is devoted to explanations of these physical and physiological properties. If you have little experience with these properties, we recommend that you work through this chapter first. If you have more experience, this chapter can be used as a reference.

Designers will find interesting and unusual color creations in this book, colorists in color laboratories can find inspiration for analyzing car paints, and anyone interested in these "crazy" pigments will find this book a treasure trove of visual highlights.

All brand names are the property of their respective manufacturers. The results shown refer to my sample panels and cannot be used as delivery conditions.

The focus is on color measurement of interference pigments. Only this allows for the precise description, characterization, and identification of the color and color gradient of an interference pigment. Only in this way could their properties also be obtained in blends and applications.

2 Fundamentals

The path to color

The path to color can be divided into the following sections: To our eyes, all processes are physical in nature. White light is manipulated in different ways by pigments and dyes. When it hits the retina in the eye, these processes, as well as those in the brain, are described by physiology. An optical stimulus is triggered in the retina, which is transmitted to the brain, where it produces a color reaction. Psychological processes also take place in the brain, accompanying the "evaluation" (Figure 7).

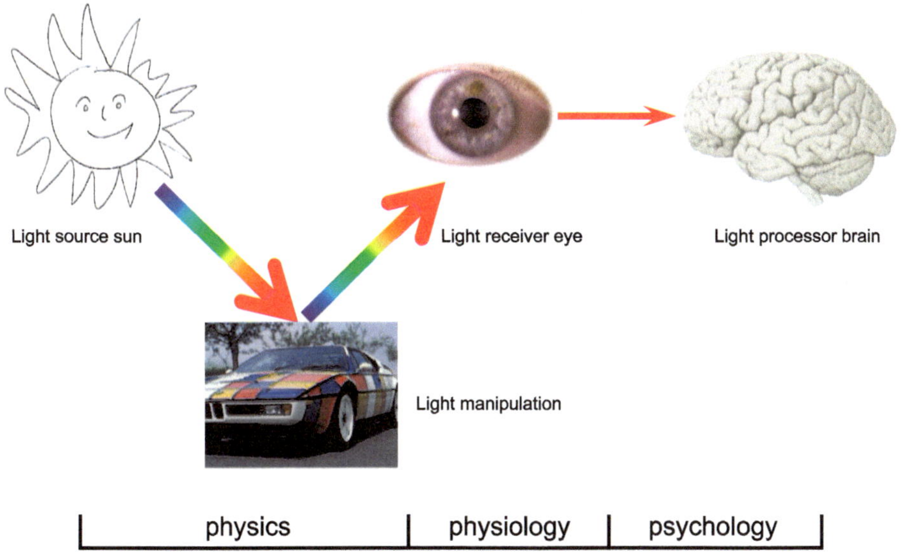

Figure 7: From the sun to the eye to the brain: How colors are created.

Before our eye

When light hits pigments or dyes, it is manipulated by absorption, reflection, or interference. The fact that no colors are produced here becomes clear when the light changes. If, for example, we take sample panels in the four colors yellow, red, blue, and green, we only perceive them as such in white light (Figure 8). If we illuminate the same panels with red light, the blue and green panels appear black because they do not reflect red light in their spectral range (Figure 9). With turquoise light, yellow becomes green-yellow also reflects in the green and red ranges, which are not reflected by turquoise light. Red appears as black, blue as blue, and green as light blue because its yellow component is not reflected (Figure 10).

https://doi.org/10.1515/9783111439204-002

Figure 8: We recognize colors with white light.

Figure 9: Red light makes green and blue appear black.

Figure 10: With turquoise light, red becomes black and yellow becomes yellow-green.

If violet light is used for illumination, yellow and red are reproduced as red, and green and blue as blue (Figure 11).

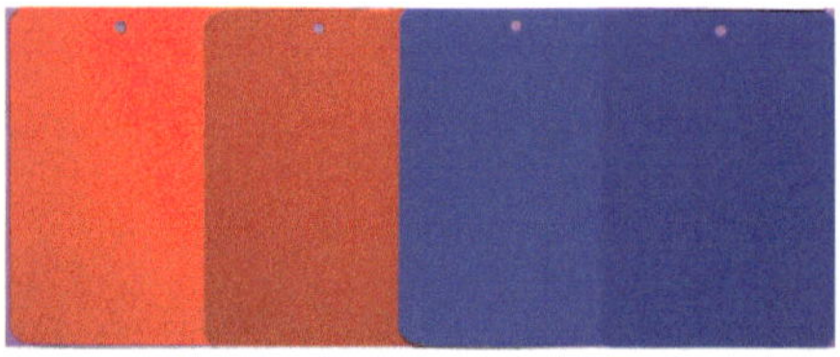

Figure 11: Purple light turns yellow into red and green into blue.

The light manipulated and reflected by the pigments and dyes is displayed and measured in the visible spectral range from 400 to 700 nm. In this spectral range, we perceive colors ranging from blue-violet at the shortest wavelengths to blue, green, yellow, and orange to red at the longest wavelengths (Figure 12). The spectral range can also be represented over frequencies from $750 * 10^{12}$ to $375 * 10^{12}$ Hertz.

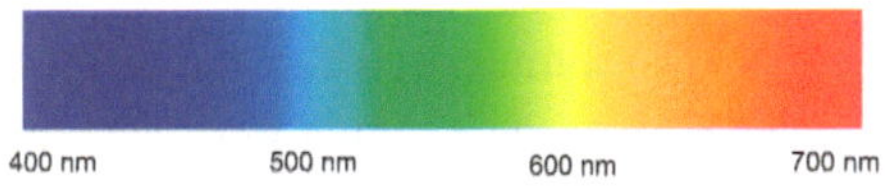

Figure 12: The visible spectrum ranges from blue-violet to blue, green, yellow, orange, and red.

The reflections are usually measured at intervals of 10 nm (1 nm = 1 nanometer = 10^{-9} m) and shown as a curve in a diagram. The reflection curves show the reflections as a percentage compared to a white standard, which is used to calibrate the corresponding measuring device beforehand (in Figure 13 as a measurement curve, in Figure 14 shown as a bar chart). Some charts also start at 330 nm to describe the optical properties in the near UV range. These measurements were performed with the ZEISS GK311/M. For

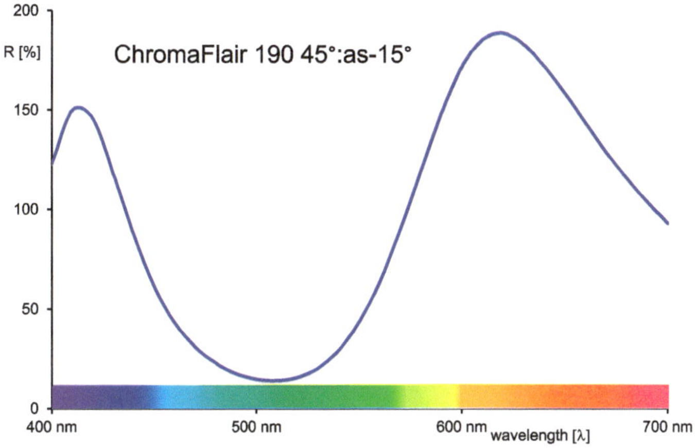

Figure 13: A reflection curve reflects the reflections in the spectral range.

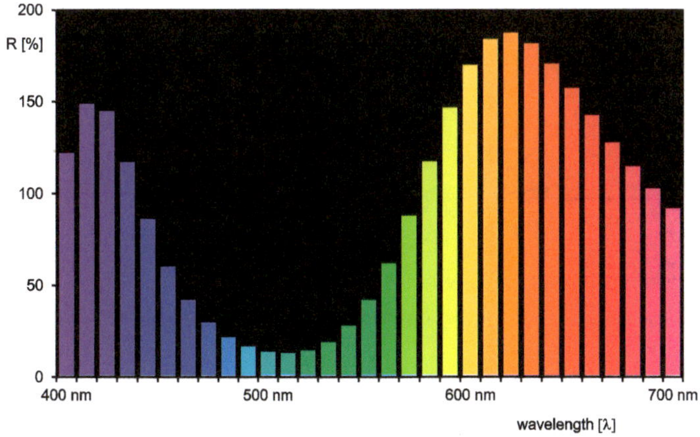

Figure 14: This illustration clearly shows the reflected colors.

all measurements of interference pigments, the corresponding geometry is specified, which consists of the illumination and observation angle or the illumination angle and the difference angle to the corresponding specular angle, for example, 45° illumination and –30° observation (= 45°:–30°) or 45° illumination and 15° difference angle to the gloss (= 45°:as15°, as = aspecular). This is the only way to reproduce an exact representation of the effects (Figure 15).

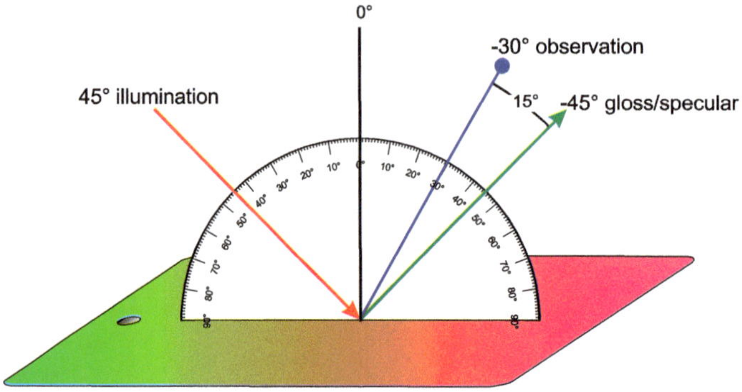

Figure 15: Geometry is composed of the angles of illumination, gloss, and observation.

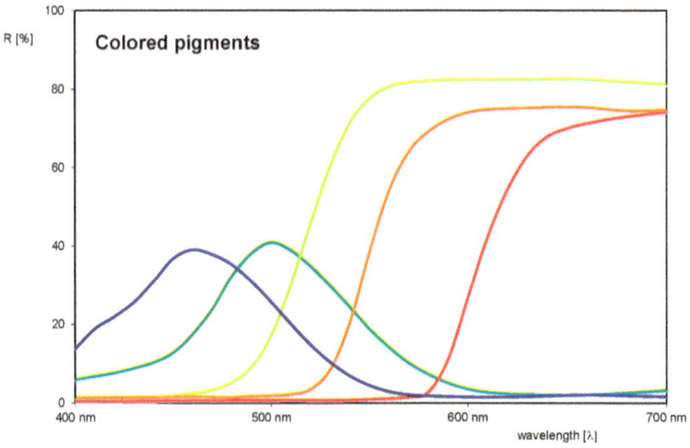

Figure 16: Blue and green pigments have a maximum reflectance, while yellow and red pigments have a reflectance plateau.

Colored pigments exhibit typical reflections: green and blue pigments each have a maximum in their spectral range, while red and yellow pigments exhibit a reflection plateau (Figure 16). Interference pigments, on the other hand, have long rising and falling reflection curves, as shown by a comparison of a yellow interference pigment with a yellow color pigment (Figure 17). In most cases, the reflections of interference pigments close to the gloss are significantly higher than those of colored pigments.

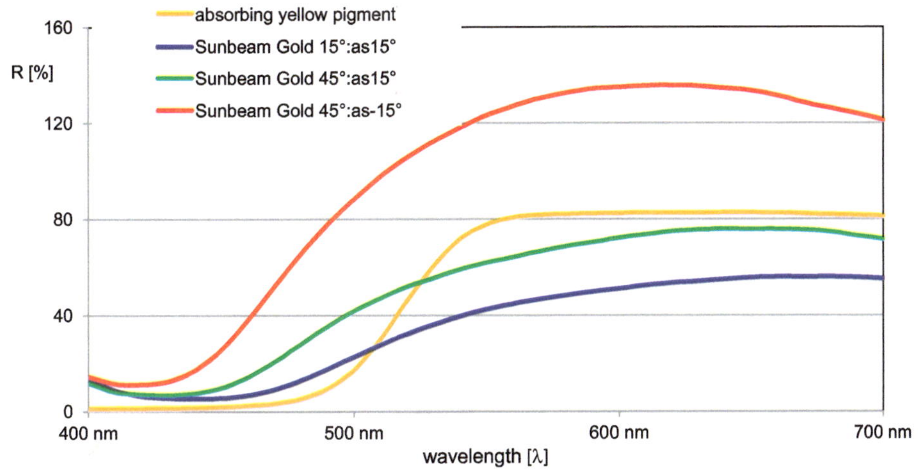

Figure 17: Interference pigments show long reflection increases compared to colored pigments.

Behind our eye

When light rays enter the eye and reach the retina, they trigger an optical stimulus in the three-color cones for red, green, and blue, which is transmitted to the brain via the optic nerve. The yellow spot on the retina is the area of sharp vision, while the blind spot is where the optic nerve leaves the interior of the eye. The lens of the eye acts as a simple optical system, allowing the sharpness of the image to be adjusted. Blue light rays are refracted more strongly towards the center of the image than red rays, a phenomenon known as chromatic aberration.

The abnormal development of a color cone leads to color vision deficiency, also known as color blindness. The most common types of color vision deficiency involve the red and green cones. Red color vision deficiency is called protanopia, while green color vision deficiency is called deuteranopia. In both cases, depending on the severity of the color vision deficiency, red and green colors cannot be recognized or distinguished. The illustration shows a van with a red body and a green driver's door. A person with color vision deficiency sees both colors as brown ochre and cannot distinguish between them (Figure 18).

The visual stimulus leaves the retina and is transmitted to the brain via the optic nerve. There it is converted into colors. The greatest achievement of our brain is the combination of light rays at the respective ends of the color spectrum, blue-violet and

Figure 18: People with color blindness cannot tell the difference between the driver's door and the body.

Figure 19: The brain combines the end colors of the color spectrum to form purple.

red, to form purple (Figure 19). This color does not correspond to a single wavelength in the visible spectrum but is composed of blue-violet and red-light rays. This combination creates the familiar color wheel, in which one color transitions into the next (Figure 20). A purple pigment also exhibits a combination of red and blue reflections (Figure 21).

Figure 20: The combination of the end colors leads to the color wheel.

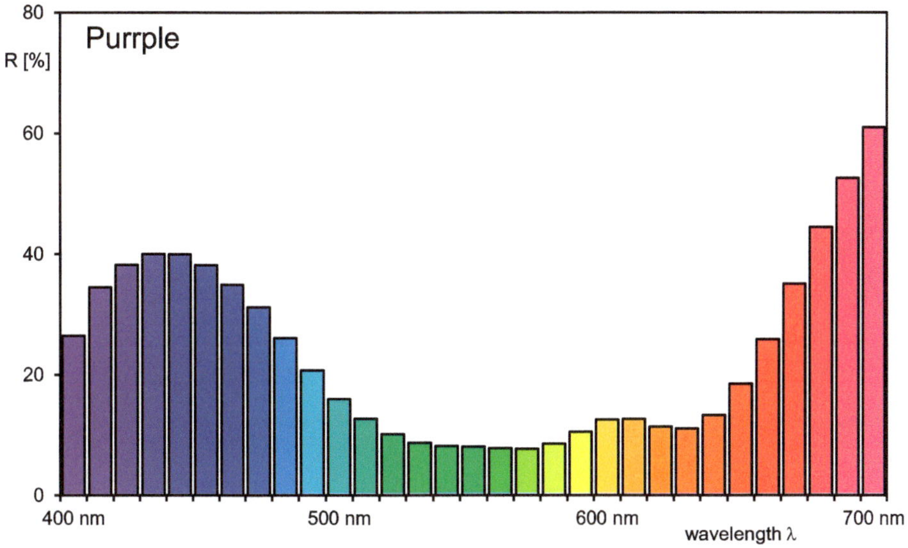

Figure 21: A purple pigment reflects in the blue-violet and red spectral range.

Two characteristics of our brain are important for industrial applications: Our brain cannot "store" color, i. e., I cannot view a color in one place and reproduce it in another. The brain is very good at recognizing color differences, which is important for work in a color laboratory.

The color values are represented using an a*b* diagram: No blue can be yellowish at the same time, but rather greenish or reddish. And no yellow can be bluish, but only greenish or reddish.

Red cannot be greenish, only yellowish or bluish. And green cannot be reddish, only yellowish or bluish. For this reason, the color values are represented in an a*b* diagram, where the x-axis represents the proportion of red and green. Positive values represent the red component +a*, and negative values represent green with –a*. Yellow components are shown on the y-axis with positive b* values and negative values for the blue component with –b* (Figure 22).

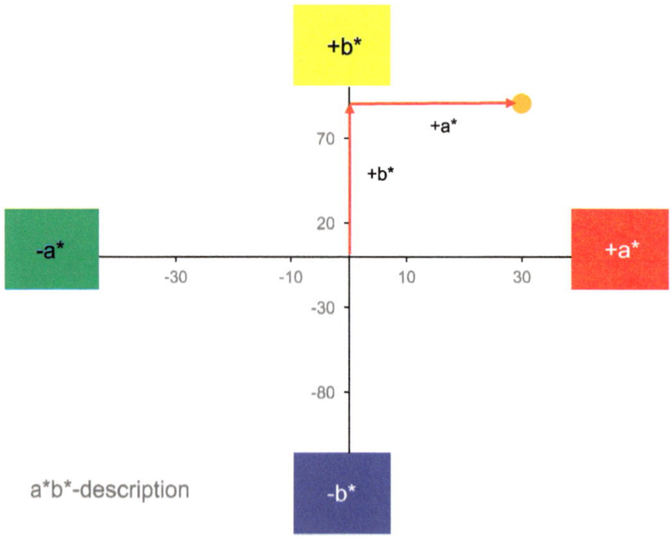

Figure 22: In the a*b* diagram, a color is represented by its a* and b* values.

A color value can be described using the specification a*b* with the respective proportions of red-green and yellow-blue. Another description describes a color in the a*b* diagram using the chroma C*, which corresponds to the distance of the color value from the zero point of the a*b* coordinate system. The larger this value is and the greater the distance from the zero point, the more colorful the color is. The hue angle h° describes the location that is removed from the red axis in degrees: 90° corresponds to the yellow axis +b*, 180° to the green axis –a*, and 270° to the blue axis –b* (Figure 23). The brightness values L* and chroma values C* are shown in bar charts. All values are dimensionless.

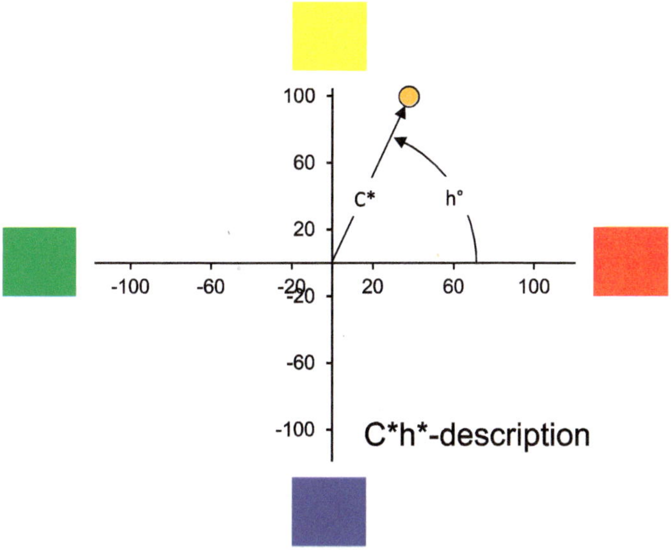

Figure 23: Color can also be described by its chroma C* and hue angle.

Geometries

The colors and effects of interference pigments depend on the angle of illumination and the angle of observation. The angles are usually specified based on the normal, which is perpendicular to the object. Although the optical law “angle of incidence = angle of reflection” applies, the angles are specified with positive and negative signs. As a rule, positive angles are on the side of the illumination (Figure 24).

A second way of representing angles is the protractor method. This involves placing a semicircle over the object, starting at 0° and ending at 180° for the normal at 90° (Figure 25).

The geometries – combination of angles – are specified in the following form: 45°:as15° (illumination 45°:aspecular 15° = angle from the corresponding specular); this representation is also possible: 45°/as15° (illumination angle/difference angle from gloss). Illumination and observation angles can also be specified, for example 45°:–30° or 15°:0°. Both denote an aspecular angle of 15° from the gloss angle at –45° or –15°, respectively.

Negative difference angles (aspecular) are on the side of the gloss angle opposite the illumination angle (= trans position). Positive difference angles are on the same side of the gloss angle as the illumination (= cis position).

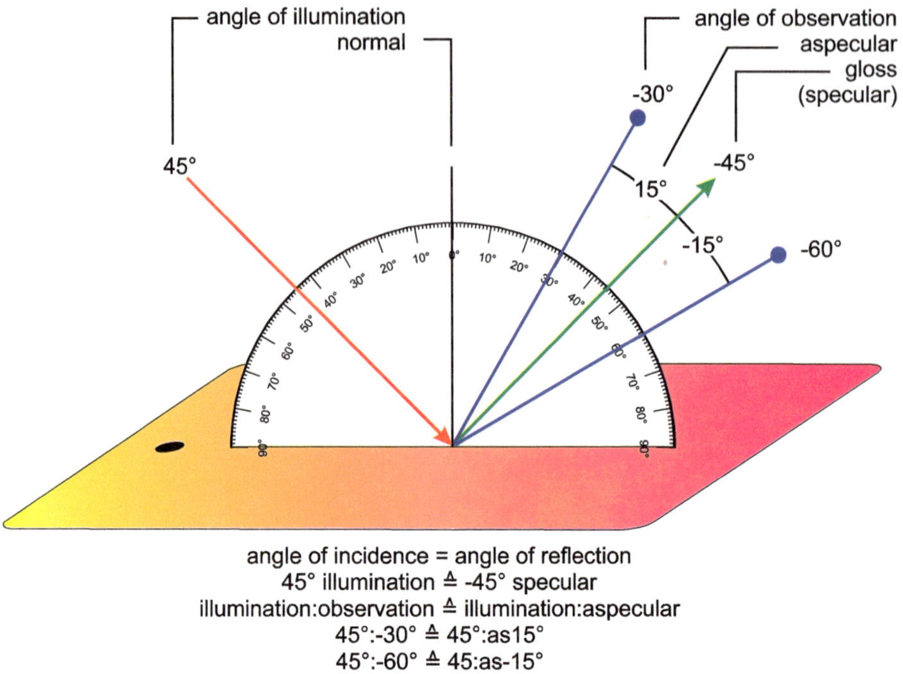

Figure 24: A geometry describes the angles of illumination, gloss, and the difference angle from the gloss.

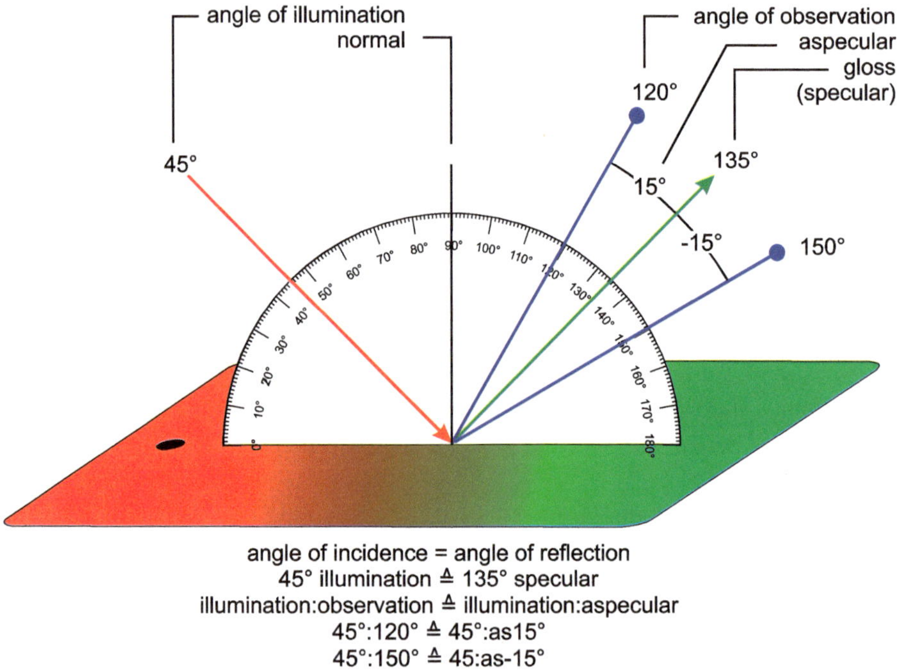

Figure 25: In the Proctrator representation, the semicircle runs from 0° to 90° (normal) to 180°.

All illustrations in this book are based on painted panels that have been created over the past almost 40 years. Interference pigments from various pigment manufacturers were selected for this purpose and incorporated into appropriate paint systems. In most cases, they were applied in a two-coat process with a base coat containing pigments and a clear coat to seal the base coat. They were then measured colorimetrically using measuring devices from various manufacturers and evaluated.

3 Interference

Light waves are described by their wavelengths. Waves are created by circular movements, with the wave reaching its maximum at 90°, passing through zero at 180° and reaching its minimum at 270°. At 360°, it returns to zero. The distance between 0° and 360° is referred to as the wavelength.

The color spectrum describes how our brain reacts to colors based on wavelength: blue-violet and blue colors are based on light waves with short wavelengths (short wavelength >400 nm), while red colors represent long wavelengths (long wavelength <700 nm). Medium-length light waves produce green colors in the brain. Yellow colors are created by combining green, yellow, and red-light waves.

Modern interference pigments are based on transparent or nontransparent carrier plates coated with highly refractive material. The carrier plates of the most used interference pigments consist of natural or artificial mica, silicon dioxide (silica), borosilicate, or aluminum oxide (alumina), which are coated with titanium dioxide in a wet chemical process. Alternatively, iron oxide or a combination of iron oxide and titanium dioxide is used. The former version produces reddish-brown interference pigments, while the latter produces golden-yellow interference pigments. The combination of chromium oxide and titanium oxide produces green interference pigments. These pigments with iron oxide or chromium oxide are less transparent than interference pigments with titanium dioxide.

Interference pigments produced in a high vacuum have a nontransparent aluminum core. This is coated with highly refractive materials.

For automotive use, all pigments are treated and additionally protected against weathering.

Principle of interference

The reaction of interference pigments can best be described using transparent interference pigments based on mica coated with titanium dioxide. The simple interference formula explains the reaction of the pigments to incident light and shows the dependence on the layer thickness of the coating material, its refractive index, and the angle of the incident light.

The wet chemical manufacturing process for a blue interference pigment, for example, mainly results in blue pigments, and to a lesser extent also in yellow, red, and green pigments. A new batch is often color-matched to previous interference pigments.

When white light strikes a transparent interference pigment at an angle α, in which, for example, mica or alumina is coated with titanium dioxide as a carrier material, it is partially reflected at the surface (A). The remaining light rays pass through the titanium dioxide layer, refracting at the perpendicular (A–B). At the boundary layer to the carrier

https://doi.org/10.1515/9783111439204-003

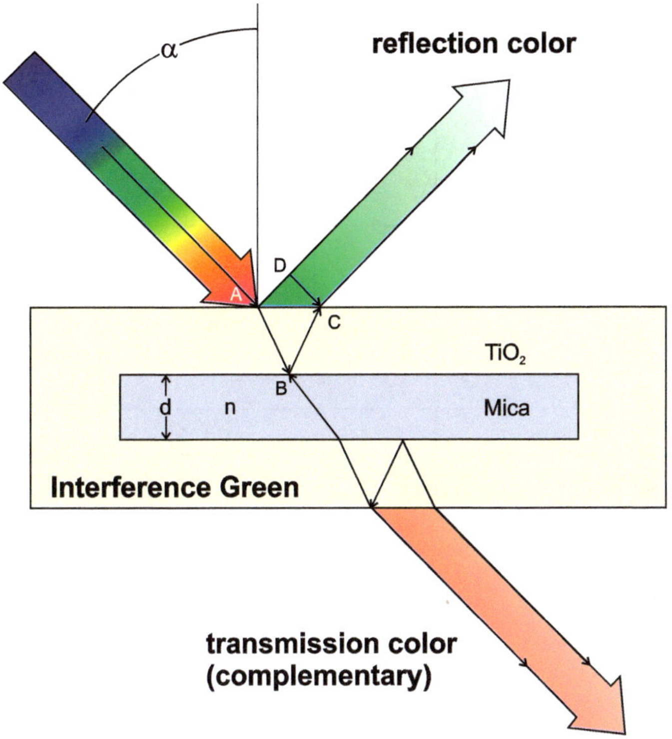

$$\delta = n(AB + BC) - AD$$

$$\Delta = 2d\sqrt{n^2 - \sin^2\alpha} + \frac{\lambda}{2}$$

d = thickness of layer
n = refractive index
α = angle of illumination

Figure 26: Reflection and transmission color.

material, part of the light is reflected again (B – C), leaving the pigment parallel to the first part at C (Figure 26).

Compared to the first part, there is a shift in the light waves due to the optical path length:

$$d = n(\mathrm{AB} + \mathrm{BC}) - \mathrm{AD}$$

$$d = 2n\mathrm{AB} - \mathrm{AD}$$

$$\Delta = 2d\sqrt{n^2 - \sin^2\alpha} + \frac{\lambda}{2}$$

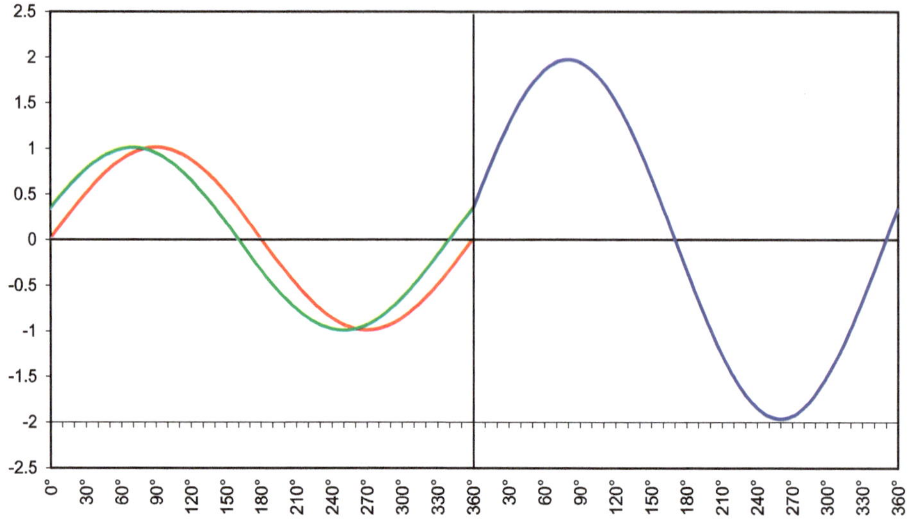

Figure 27: Light waves are described as sine waves that can add up: on the left, two light waves; on the right, the resulting wave.

The optical path length depends on the layer thickness d of the titanium dioxide, its refractive index n, and the angle of incidence α of the light rays. Since there is a transition from an optically thinner to an optically denser medium, an additional phase shift of $\lambda/2$ occurs.

Both reflected parts interfere with each other, i. e., if a wave crest encounters another wave crest, this leads to an amplification of the resulting wave (Figure 27). In the ideal case, a trough on a wave leads to the cancellation of the resulting wave, but in this case, it leads to a reduction (Figure 28). For example, if the waves of the green light rays of both reflected parts are shifted so that maximum meets maximum, they are amplified. The waves of other light rays – for example, red and yellow – are then shifted relative to each other so that the resulting waves are reduced. The overall result is a green color impression.

The resulting color depends on the thickness of the titanium dioxide layer. From the white interference pigment at low layer thickness, the resulting color shifts to the longer wavelength. Measurements of the reflections on mica-based interference pigments show this shift from minima and maxima from the UV range to the visible range. White is followed by yellow, then red, purple, blue, and then green (Figure 29).

The same thing happens with interference pigments based on alumina (Figure 30). The layer thickness is determined during the manufacturing process.

This shift in reflections can also be detected based on the reflection maxima of the colorful interference pigments (Figure 31).

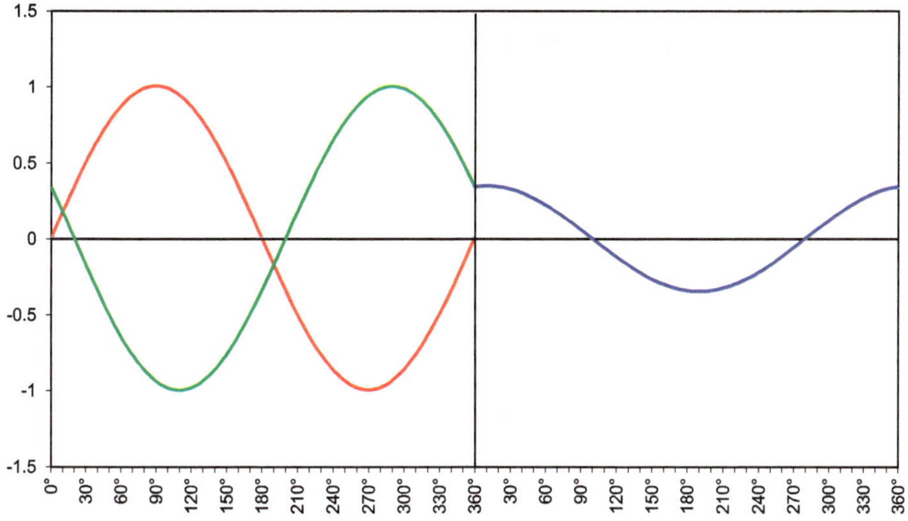

Figure 28: If two light waves are shifted against each other (left), the resulting wave becomes significantly smaller (right).

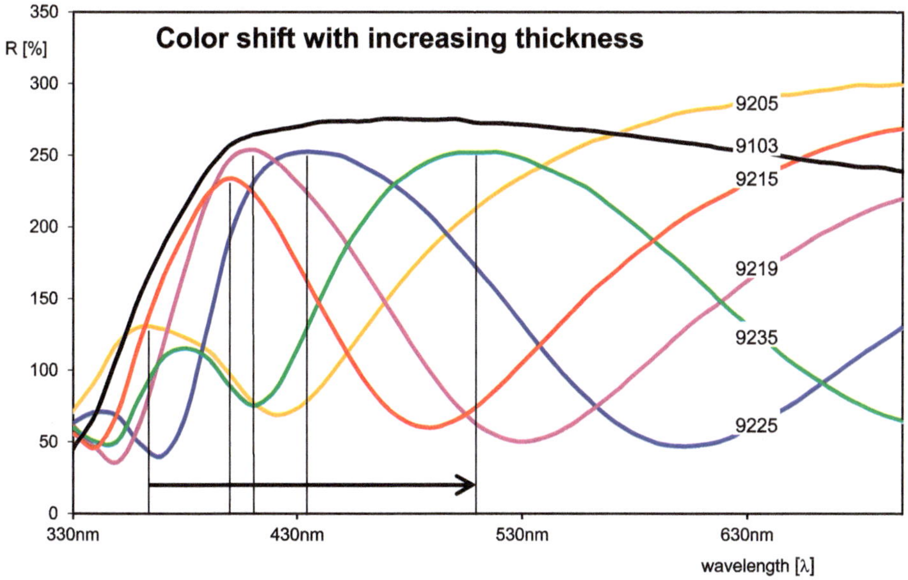

Figure 29: Mica-based interference pigments show significant color shifts toward longer wavelengths with increasing layer thickness of titanium dioxide (white, yellow, red, lilac, blue, green).

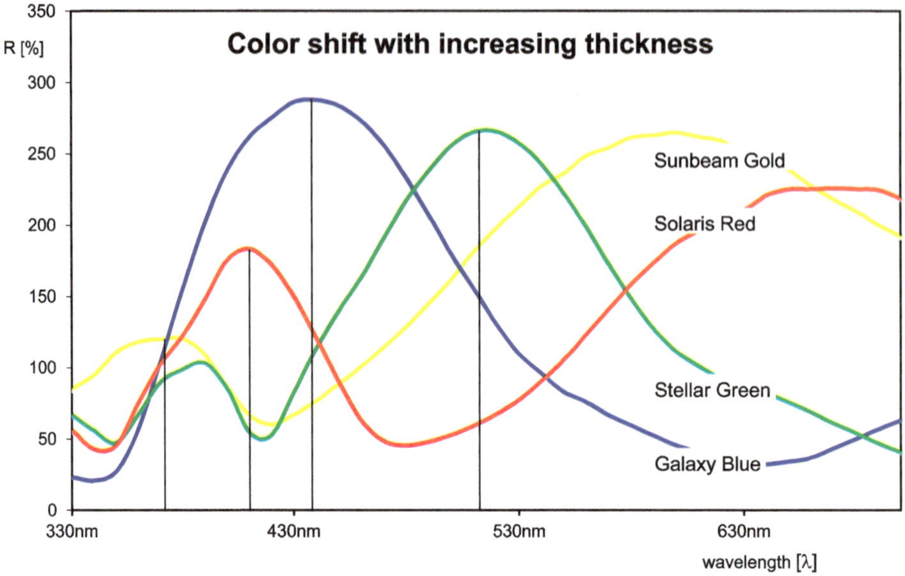

Figure 30: Alumina-based interference pigments also exhibit color shifts toward longer wavelengths.

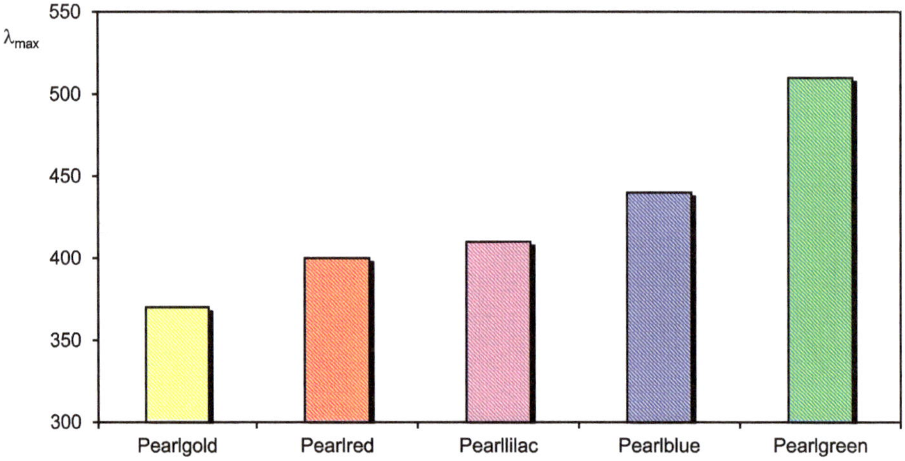

Figure 31: The reflection maximum shifts from yellow through red and blue to green.

Dependence on the angle of illumination

If the angle of illumination is changed, the optical path length also changes. If the angle of illumination becomes steeper (= smaller), the reflection shifts to longer wavelengths. If this angle becomes flatter (= larger), the reflection shifts to shorter wavelengths. If a green interference pigment (e. g., Stellar Green) is observed at a 45° illumination angle, its color shifts toward yellow when the pigment is illuminated at an angle steeper than 45°. If the same pigment is illuminated at a flatter angle, i. e., at angles greater than 45°, the green becomes more bluish (Figure 32).

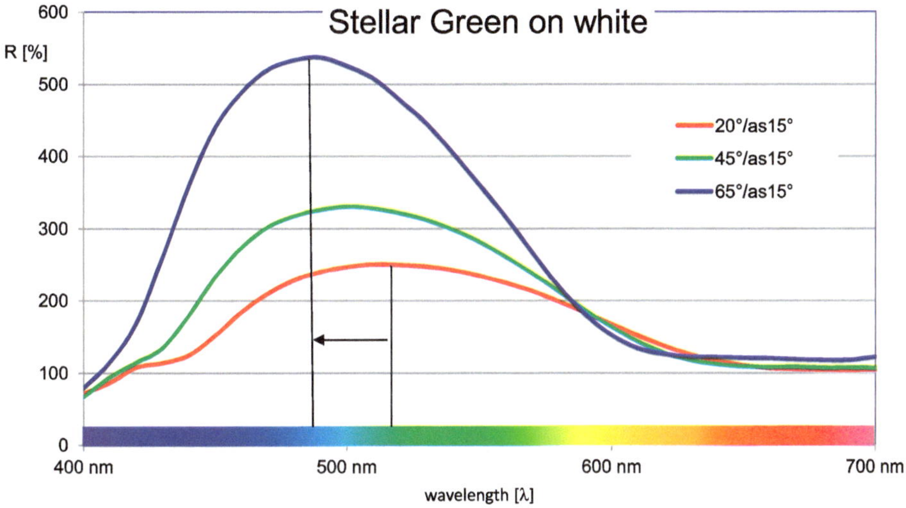

Figure 32: The reflection maxima of an interference pigment become larger and shorter in wavelength.

This color shift due to the angle of incidence of light is described by the interference line that connects the a*b* color values at steep, classic, and flat illumination angles. This shift is characteristic of a colorful interference pigment and can be used to identify it. The shift always occurs counterclockwise in the a*b* chart (Figure 33). This diagram shows the respective interference and aspecular lines of various interference pigments. The interference lines are composed of the color values at 15°:as15°, 45°:as15°, and 45°:as–15° (= 60°:as15°). The aspecular lines show the color values resulting from measurements and observations at an illumination of 45° and increasing aspecular angles (here from 45°:as15° to 45°:as110°).

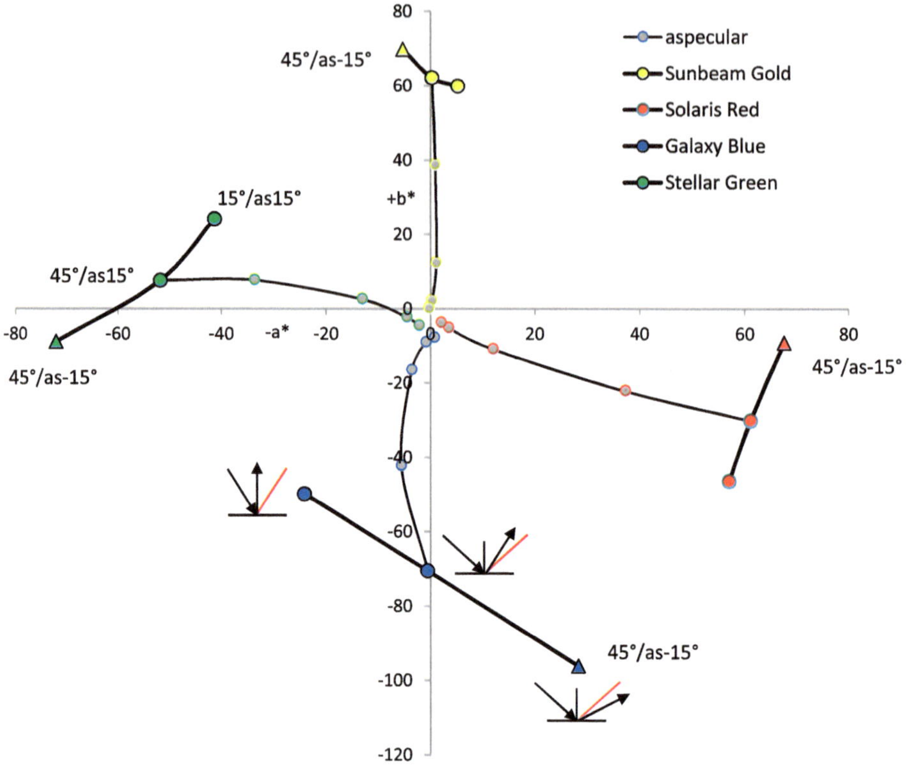

Figure 33: Typical representation of interference pigments with their interference lines from 15°:as15° to 45°:as15° to 45°:as–15° (= 60°:as15°) and specular lines from 45°:as15° to 45°:as110°.

Transmission color

The light rays that were not reflected pass through the interference pigment and exit at the rear. At the transition from the optically denser to the optically thinner medium, there is no phase shift of half a wavelength. The resulting transmission colors are therefore complementary to the reflection colors. With white interference pigments, both the reflection and transmission colors are white. However, the two whites have different origins (Figure 34).

If a transparent interference pigment such as Pearl Blue is measured "around the clock", the change from the reflection color to the transmission color can be clearly seen. With fixed illumination at 45°, this pigment was measured starting at the normal (0°, protractor 90°) and becoming increasingly flatter up to 170° (–80°, protractor 170°). From 190° (–65°, protractor 190°), the progression of the transmission color was measured (Figure 35). The sketch shows the measurement curve (Figure 36).

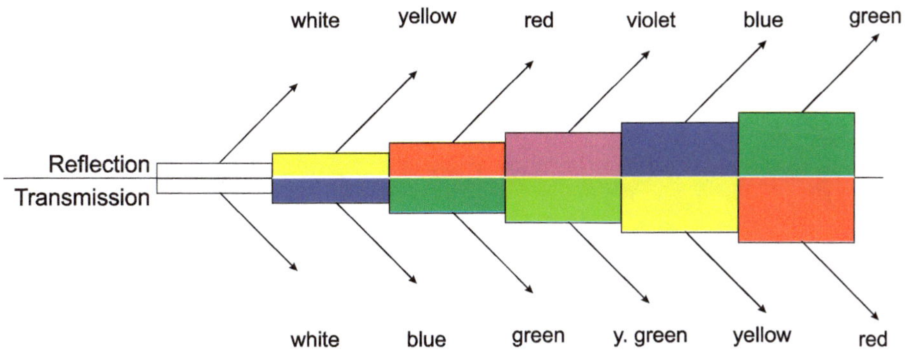

Figure 34: As the layer thickness of the titanium dioxide increases, the reflection and transmission colors of the mica pigments change.

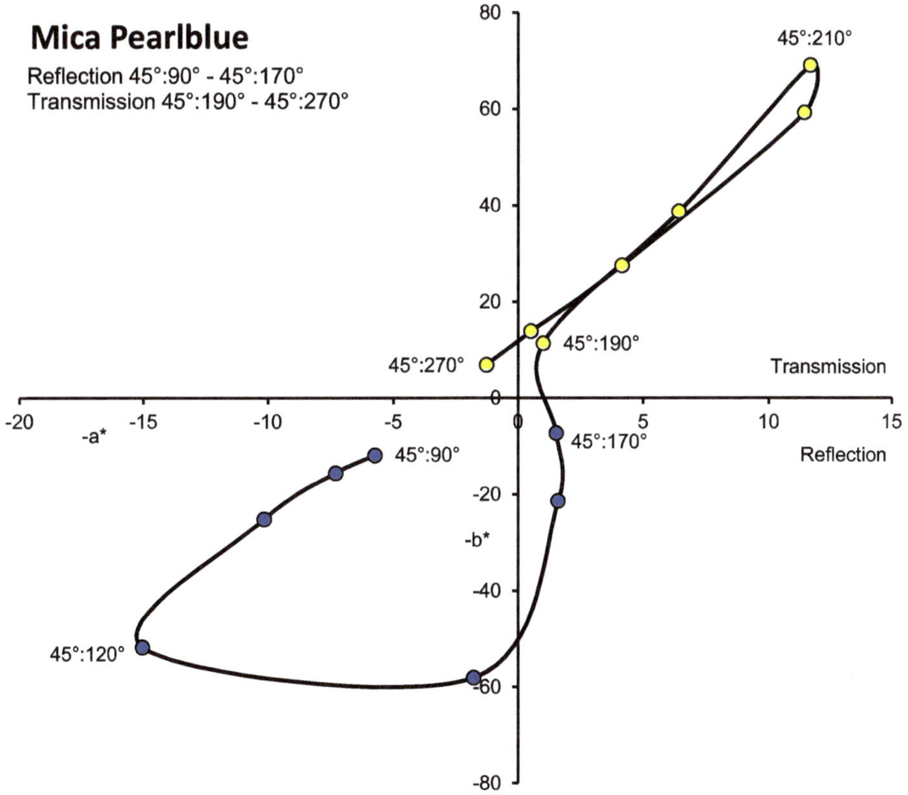

Figure 35: Perlblau was measured on transparent film.

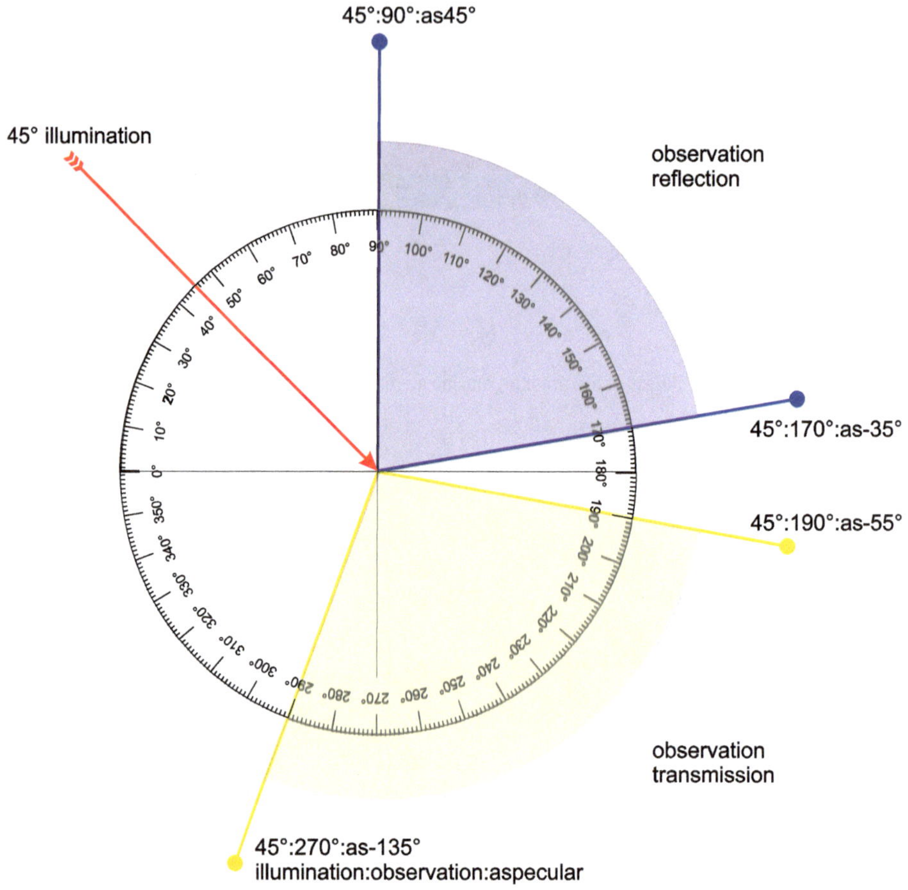

Figure 36: The schematic representation of the measurement.

Reflection curves – measured close to the gloss angle (specular) at 15° aspecular – show a color shift when the layer thickness d of the titanium dioxide increases. These are characteristic of the respective reflection color. When these pigments are measured at 25° from the gloss (aspecular), the reflection curves of the different interference colors are almost identical (Figure 37). At 45° from the gloss, they then show the corresponding transmission color when the respective interference pigment is applied to a white substrate. The example of the reflection curves of the interference pigment Stellar Green shows the change from the reflection curve at 45°:as25° to the reflection curve at 45°:as45° (Figure 38). The change from reflective to transmissive color can also be seen when an interference pigment is applied to a transparent film: when viewed from above, the reflective color is visible, and when viewed from the side, the transmission color is visible.

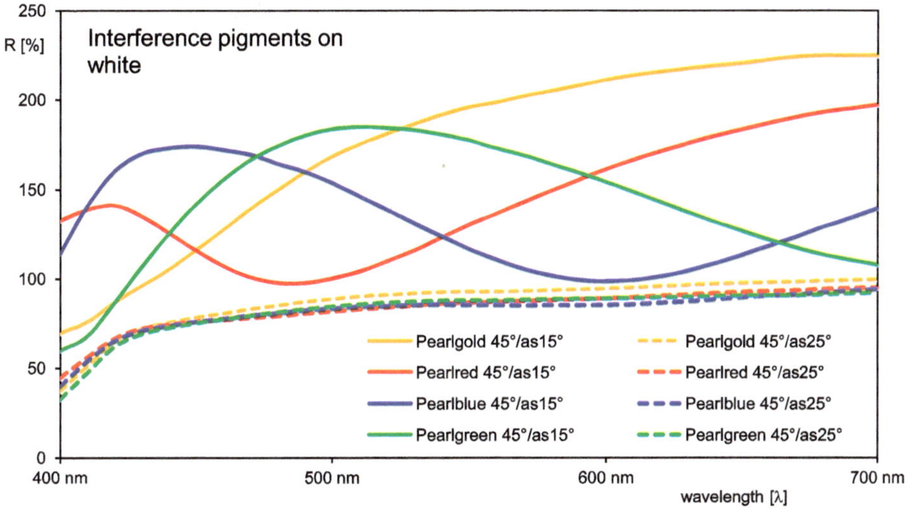

Figure 37: At 25° aspecular, the reflections from these interference pigments are virtually indistinguishable.

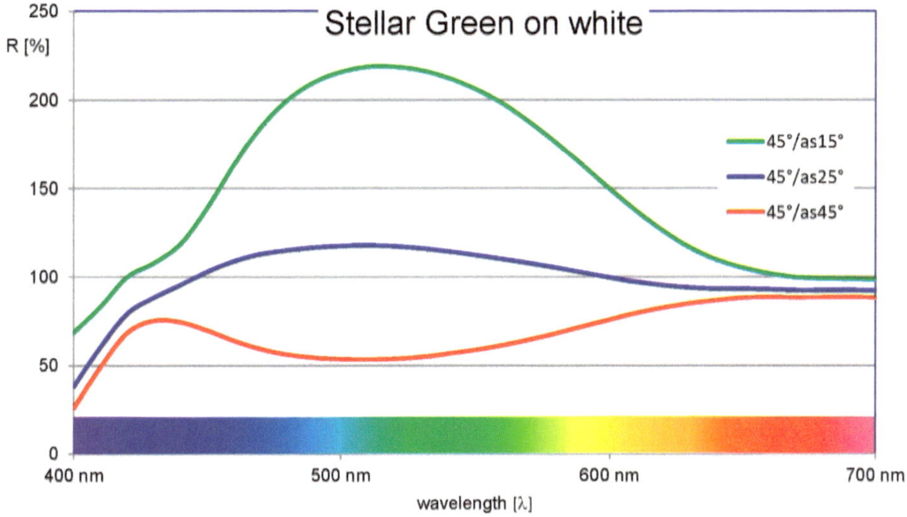

Figure 38: When transparent interference pigments are applied to a white substrate, the reflection shows the complementary color aspecular.

When interference pigments are applied to a white substrate, the representation of the a* or b* values shows the color change, which usually occurs between 20° and 30° from the gloss angle. Depending on the measuring device, this change from reflection to transmission color is represented in relation to the specular angle (Figure 39 in 5° increments with the Zeiss GK311/M, Figure 40 geometries 45°:as–15° to 45°:as110).

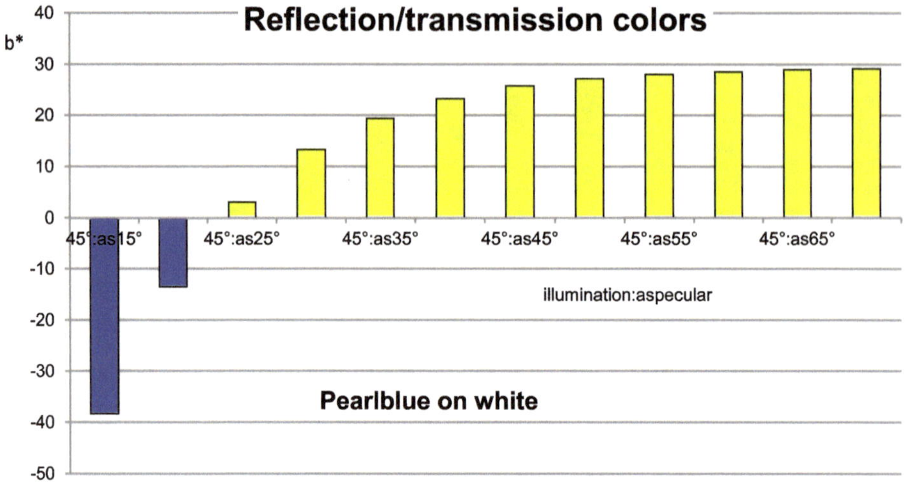

Figure 39: Transparent interference pigments have a transition area between reflection and transmission color.

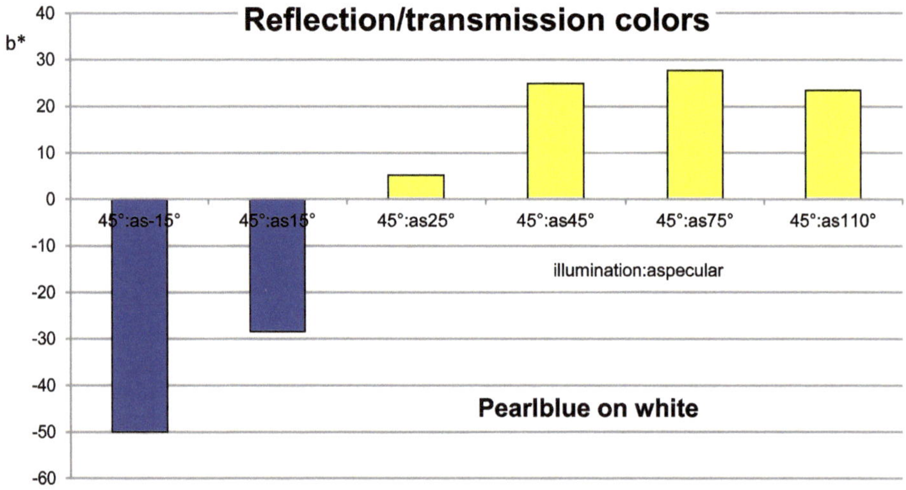

Figure 40: The transition area can also be displayed using portable devices.

Complementary interference pigments clearly show the color change when observed or measured away from the gloss (Figure 41, Figure 42). When comparing the reflection curves of 45°:as15°, 45°as25°, and 45°:as45°, you can see the difference between Pearl Blue 9225 on a white and black substrate: While the nonglossy measurements (45°:as45°) on a black substrate tend toward a neutral color, the change from the blue reflection color to the yellow transmission color is clearly visible on a white substrate (Figure 43, solid lines). Figure 44 shows the different areas, including the transition.

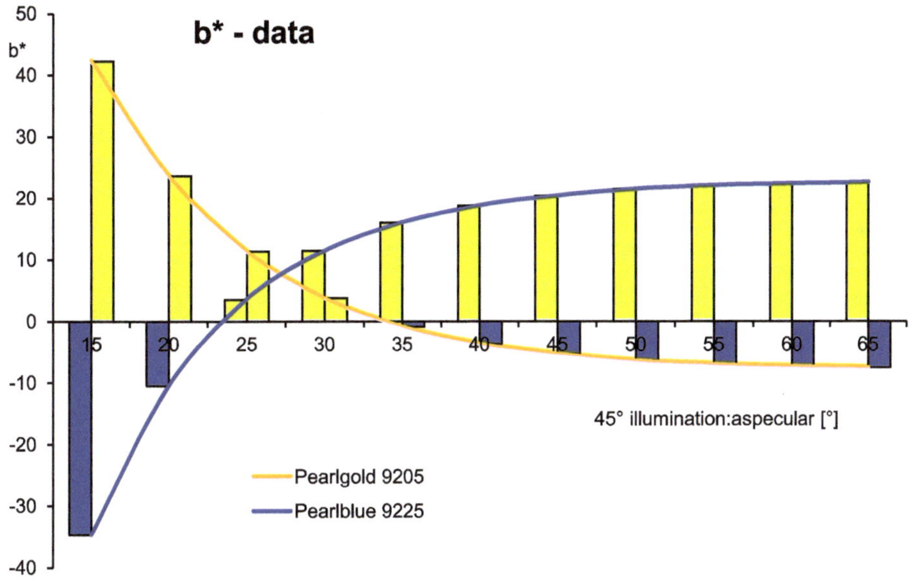

Figure 41: When displaying the b* values, Pearl Gold and Pearl Blue show their transitions to the respective transmission color.

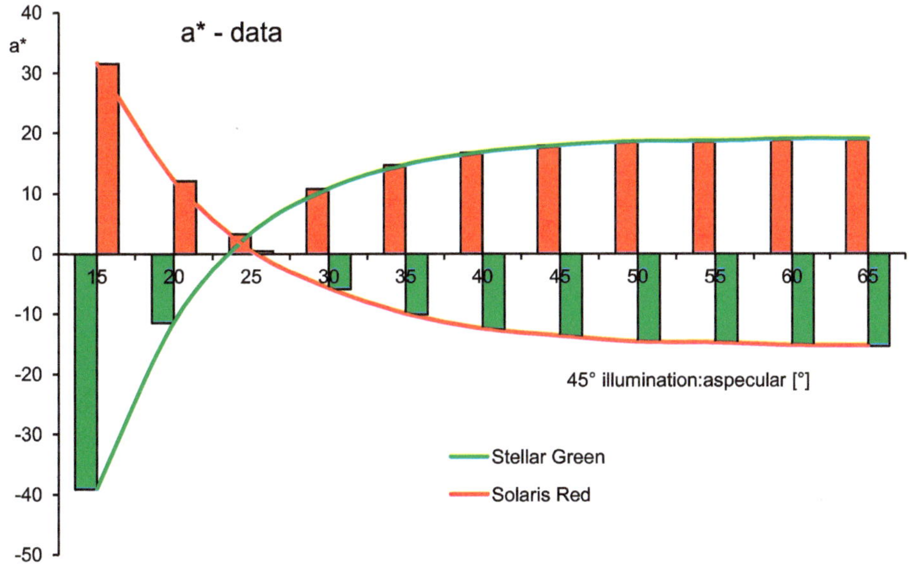

Figure 42: The color change can also be illustrated in the a* diagram for alumina-based interference pigments.

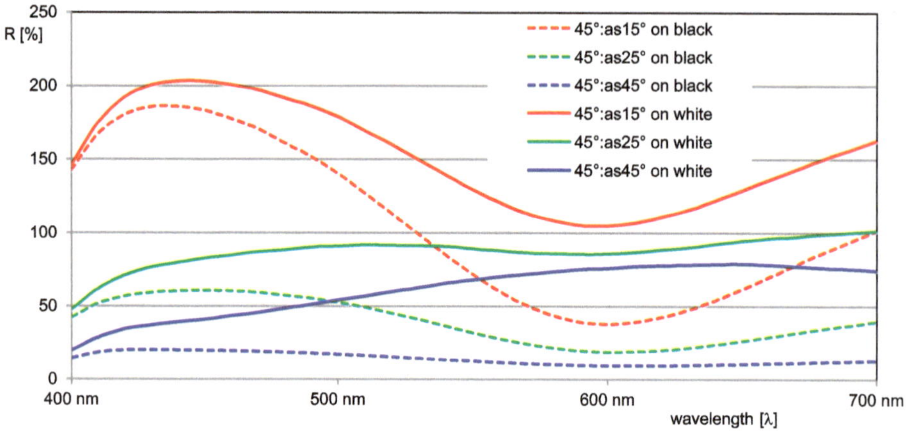

Figure 43: Different reactions on black substrate (dashed lines) and white substrate (white lines). On white substrate, the reflection color changes to the transmission color.

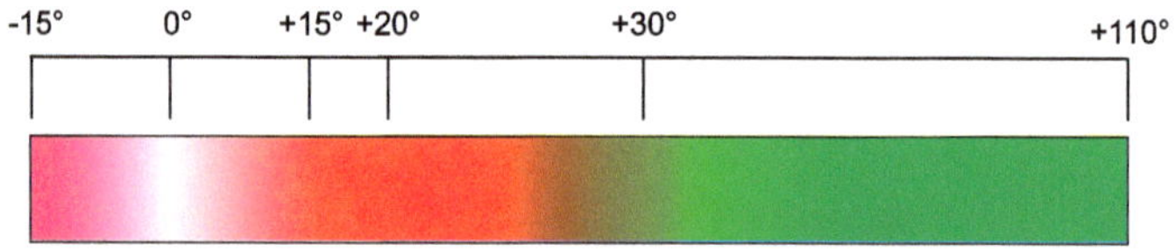

Figure 44: Schematic representation of the reflection areas with the transition area.

The evaluation of the color values C* shows comparable behavior in terms of color change. With Sunbeam Gold, this change between reflection and transmission color occurs at 30°. Here, the color value also has the lowest value. With Galaxy Blue, the change and the lowest chromaticity occur between 20° and 25° aspecular (measurements with the Zeiss GK311/M) (Figure 45, Figure 46).

The specular values and lines depend on the angle of illumination. In most cases – especially with current portable devices – illumination is at 45° and the gloss is measured at the corresponding difference angle. Other illumination angles, such as 25° or 65°, result in similar specular lines (Figure 47).

The hood in Figure 48 and Figure 49 was first painted completely white. Then the black areas were painted. Finally, stripes with various interference pigments were sprayed on. Depending on the illumination and observation, the reflective colors can be seen against the black substrate. The reflective colors can be seen on a white substrate from a certain viewing angle. If this angle is changed, the complementary transmission colors can be seen. These do not reflect off the black substrate.

Another example is the dark gray cards that have been coated with colorful interference pigments. From one angle, you can see the interference colors, while from another angle, you can see through the interference pigments to the dark gray substrate (Figure 50).

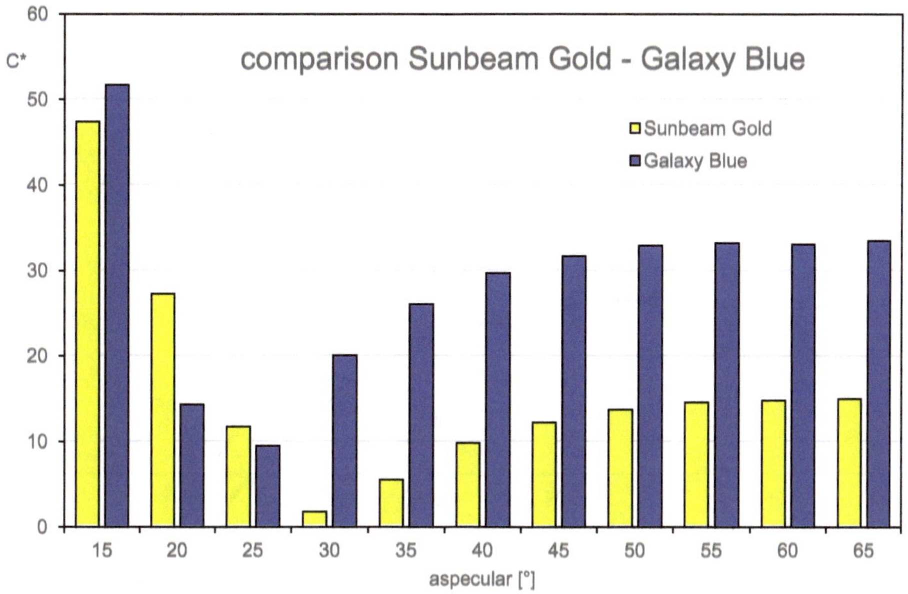

Figure 45: The chroma values C* decrease towards the transition area and then increase again. Example: Sunbeam Gold and Galaxy Blue.

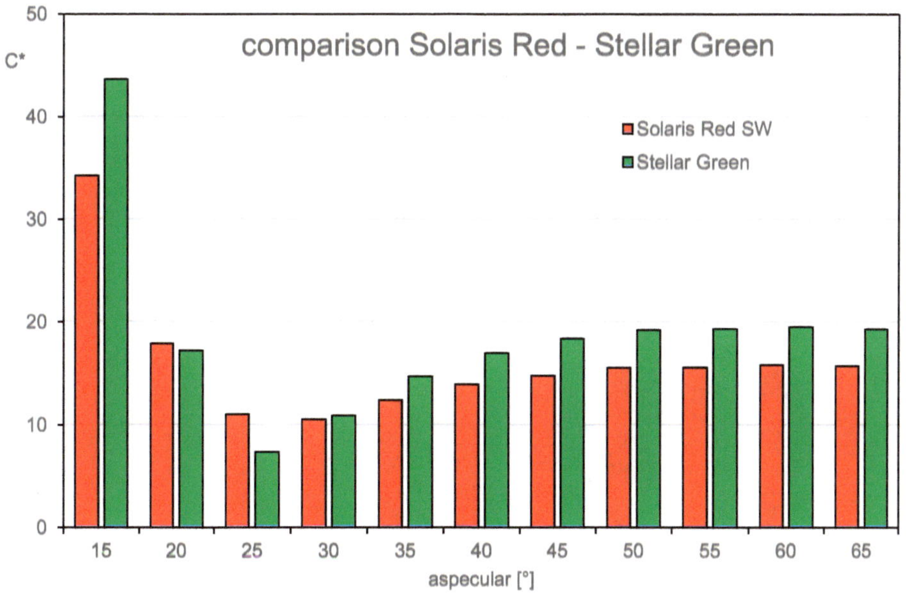

Figure 46: Color values for Solaris Red and Stellar Green.

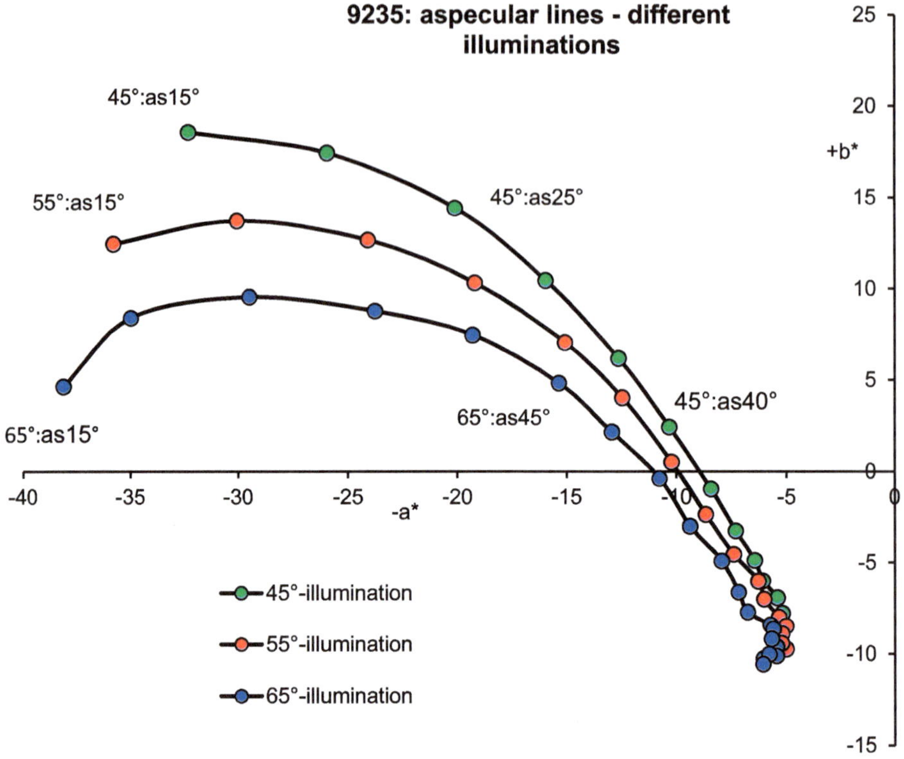

Figure 47: The specular values depend on the angle of illumination.

Figure 48: This hood shows reflection and transmission values depending on the angle of illumination and observation.

Figure 49: Transmission colors cannot be seen on a black substrate because it absorbs them.

Figure 50: From a certain angle, you can see the interference colors; from another angle, you can see through them to the anthracite-colored substrate.

Dependence on the refractive index

The refractive index in the interference formula refers to the metal oxide with which the carrier plate is coated. While titanium dioxide can be used to produce all colors of the color wheel, iron oxide only produces red to reddish-brown pigments, which are less transparent than titanium dioxide-based pigments. Combining titanium dioxide with iron oxide produces yellow to golden interference pigments. Combinations of chromium oxide and titanium dioxide result in green interference pigments. Neither the refractive index nor the layer thickness can be influenced by the user.

When the environment changes from air to water, the refractive index changes and often the viewing angle as well. The image shows water droplets on the paint surface in different colors (Figure 51). See also cover image: When light waves pass from air into water, their wavelengths become shorter. Their frequency remains the same.

Figure 51: The color of interference pigments depend on the viewing angle and refractive index: Raindrops on a fender.

Angle of illumination

The simplified interference formula also refers to the angle of the incident light. If this angle is increased, the resulting color shifts toward shorter wavelengths: the color of a yellow interference pigment shifts toward green, a green interference pigment toward blue, a blue pigment toward red, and a red pigment toward yellow. This angle can be used to determine and describe colorful interference pigments: from steep to classic and then to flat illumination, the reflection of colorful interference pigments shifts toward shorter wavelengths (Figure 52 measured from 330 to 710 nm, Figure 53 measured

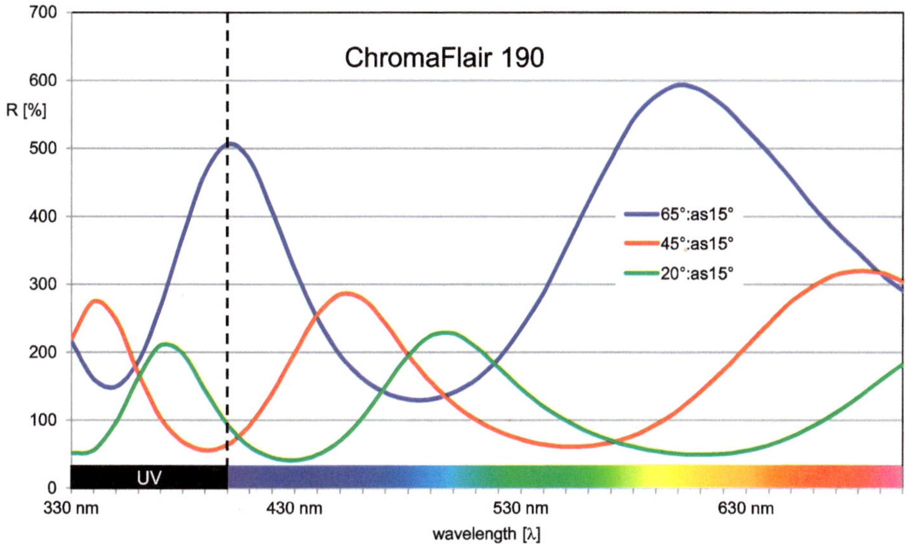

Figure 52: With ChromaFlair 190, the shift toward shorter wavelengths can also be seen in the UV range when illuminated at a flatter angle.

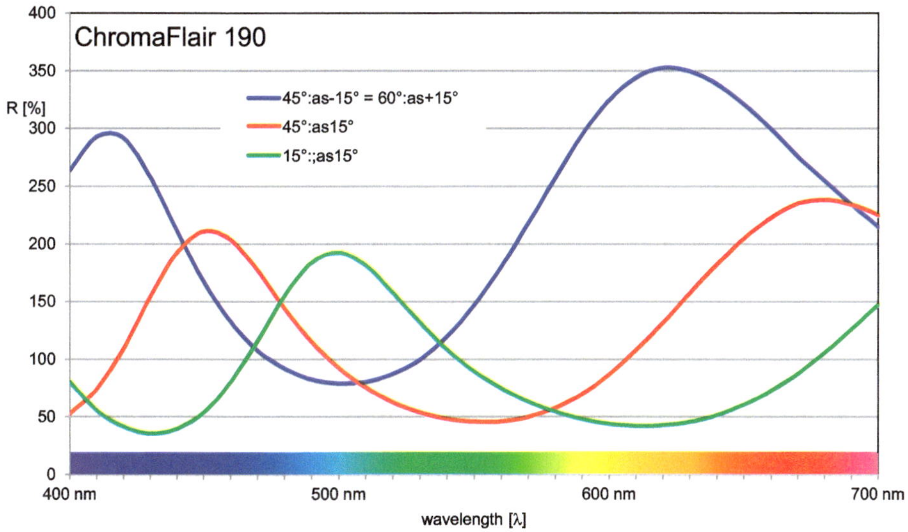

Figure 53: The portable instruments also show the shift in ChromaFlair with flatter illumination.

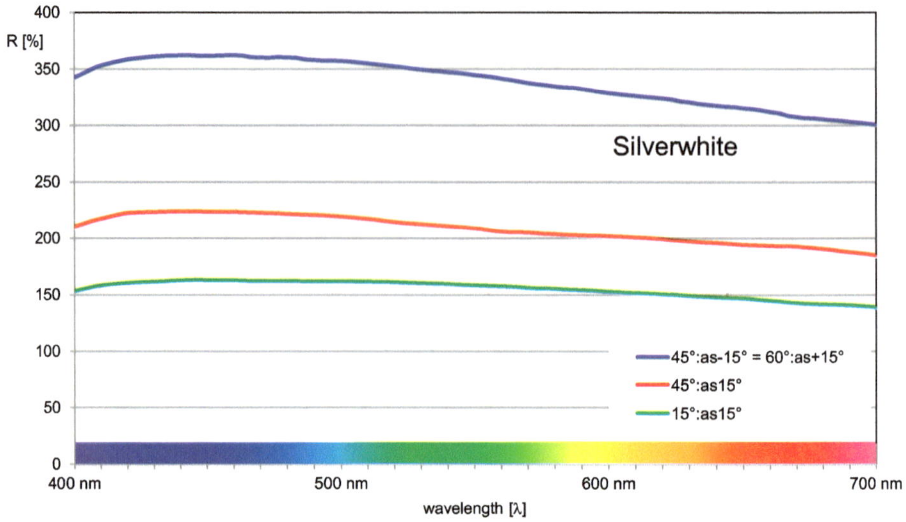

Figure 54: White interference pigments do not show any color shift, but reflection increases with flatter illumination.

from 400 to 700 nm). At the same time, the reflection increases (Fresnel's law). This phenomenon also applies to interference pigments that are produced in a high vacuum and have an aluminum core. White interference pigments behave like aluminum pigments and do not show any color shift when illuminated more flatly (Figure 54).

Interference line

The optical properties of interference pigments are represented by the interference line and the aspecular line. For the interference line, the measured values at 15° aspecular for steep, classic, and flat illumination are connected in the a*b* chart. In the standard method, ASTM E2539 (Standard Test Method for Multiangle Color Measurement of Interference Flake Pigmented Materials) only lists the two illuminations at 15° and 45°. Here, the geometry 45°:as–15° can be used as a third option instead of the flat geometry, as it corresponds to a geometry of 60°:as15° due to the law of light reversal. The diagram shows the principle of light reversal (Figure 55).

These three measurements can be used to characterize and describe any interference pigment.

With the Zeiss GK311/M laboratory device, the steepest illumination is 20° (70° protractor) at a measurement angle of –5°, resulting in an aspecular angle of 15°. The flattest illumination is 65° (25° protractor) with a measuring angle of –50° (= 15° aspecular) (Figure 56).

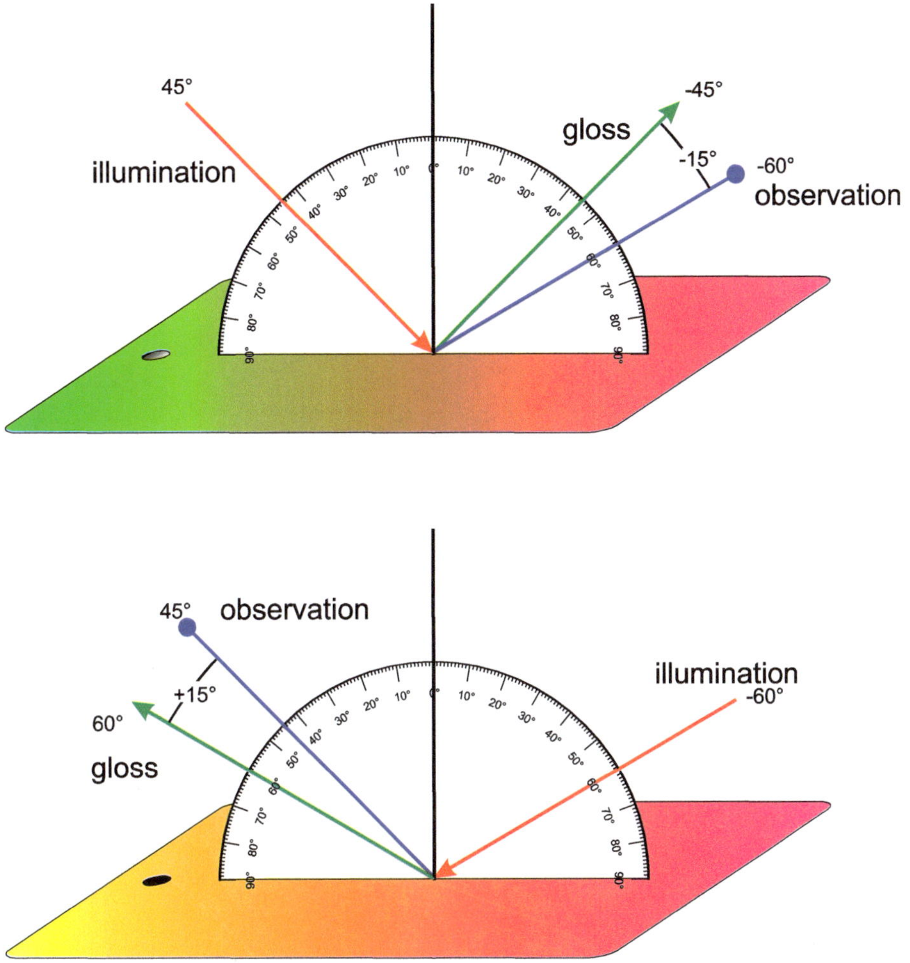

Figure 55: Due to the law of light reversal, illumination and observation can be reversed.

Interference measurement

The interference line based on the measured values for these geometries always "bends" counterclockwise for colorful interference pigments compared to the corresponding aspecular line. The aspecular line is formed from the aspecular values for the measurement geometries of 45°:as15° to 45°:as110° or 45°:as15° to 45°:as70°. White interference pigments and aluminum pigments do not "bend" away from the aspecular line but continue almost straight from 45°:as25° to 45°:as15°. They show no color shift counterclockwise (Figure 57: Aluminium + solid2, Pearl Blue + solid2).

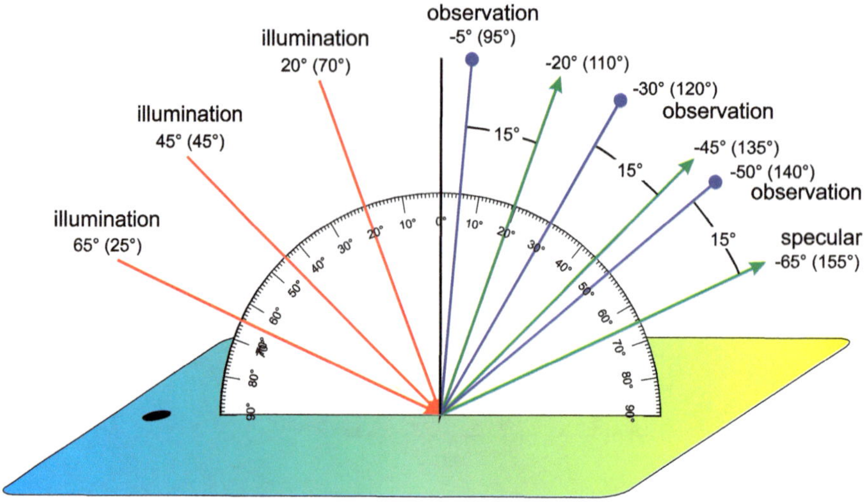

Figure 56: With the Zeiss GK311/M, illumination can be adjusted from steep to flat with corresponding measuring angles in 5° increments.

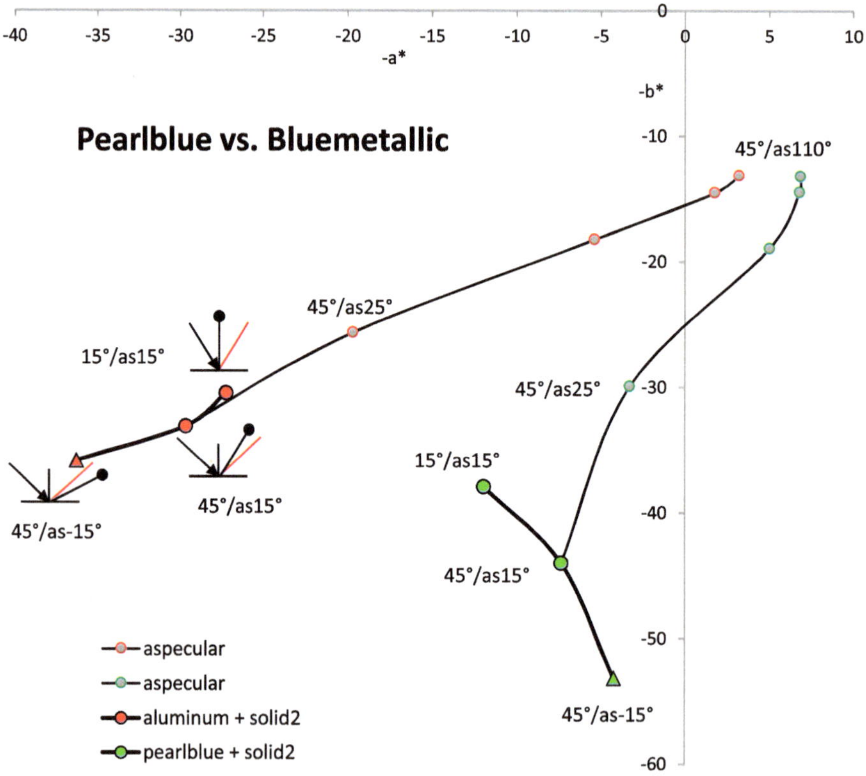

Figure 57: Aluminum and colored interference pigments can be distinguished based on their interference lines.

For the representation of the aspecular line, measurements in 5° increments at a fixed illumination angle show no particular features. Between the reflection maxima, the differences increase the closer the measurement is to the gloss. For sample sheets sealed with clear coat, it can be seen that between 5° and 10° from the gloss angle – with

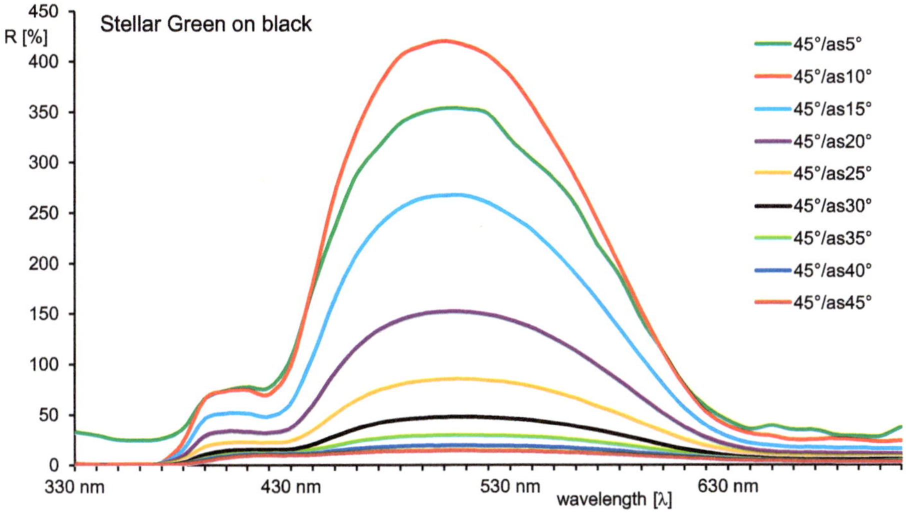

Figure 58: The closer the measurement is to the angle of specular, the greater the reflections and the greater the distance between them.

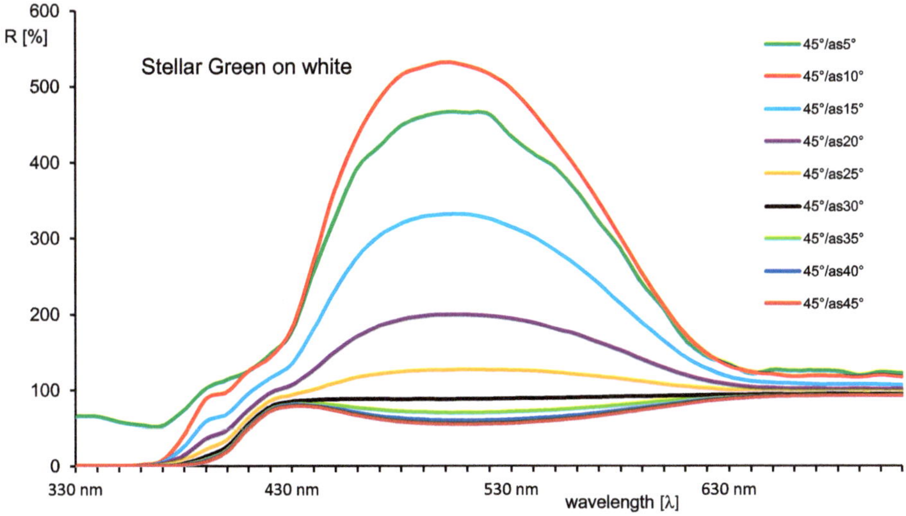

Figure 59: When measuring on a white substrate, the reflection curves reflect the change from reflection to transmission.

45° illumination – this difference is smaller than between 10° and 15°. In this respect, the measurement at 10° from the gloss angle can be classified as implausible. This applies in particular to the difference between the reflection maxima of 5° and 10° from the gloss angle (Figure 58 on black, Figure 59 on white and color change at 45°:as30°).

Transparent interference pigments exhibit a change between reflection and transmission color at angles between 20° and 30° from the gloss. This change can also be observed in blends with colored pigments or on colored substrates. Figure 60 shows the color change pattern from as–15° to as15° and in the transition area from as20° to

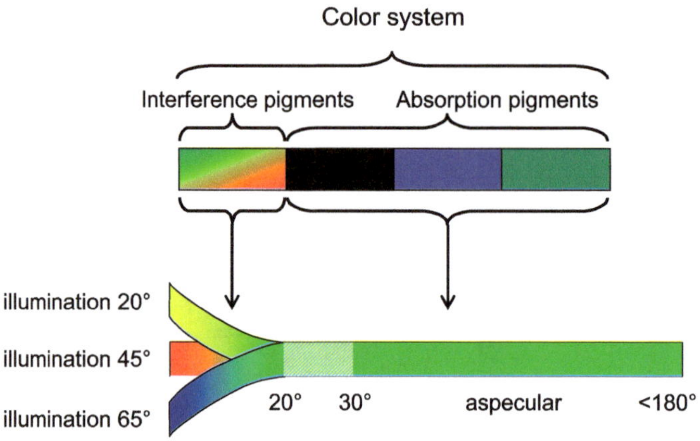

Figure 60: The color change takes place in the transition area.

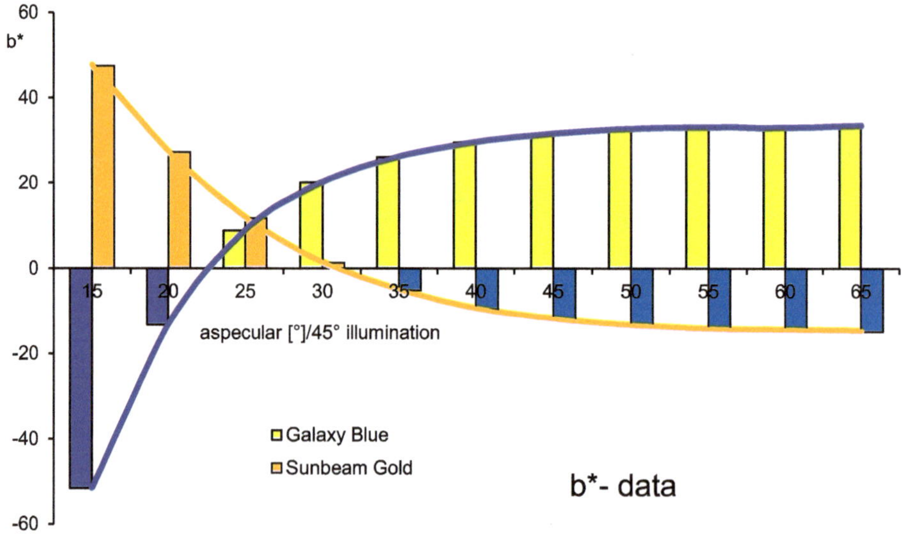

Figure 61: The change in the transition area is clearly visible in the b* diagram.

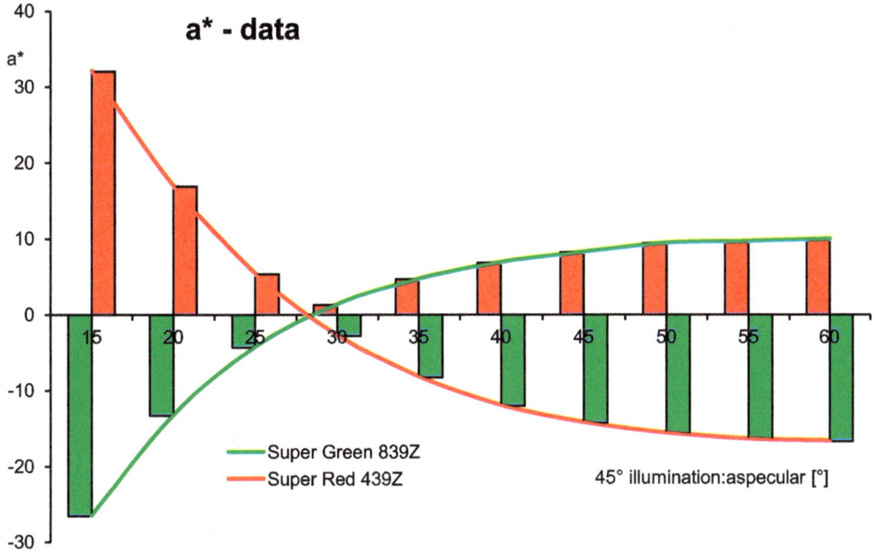

Figure 62: Green and red transitions are shown in the a* diagram.

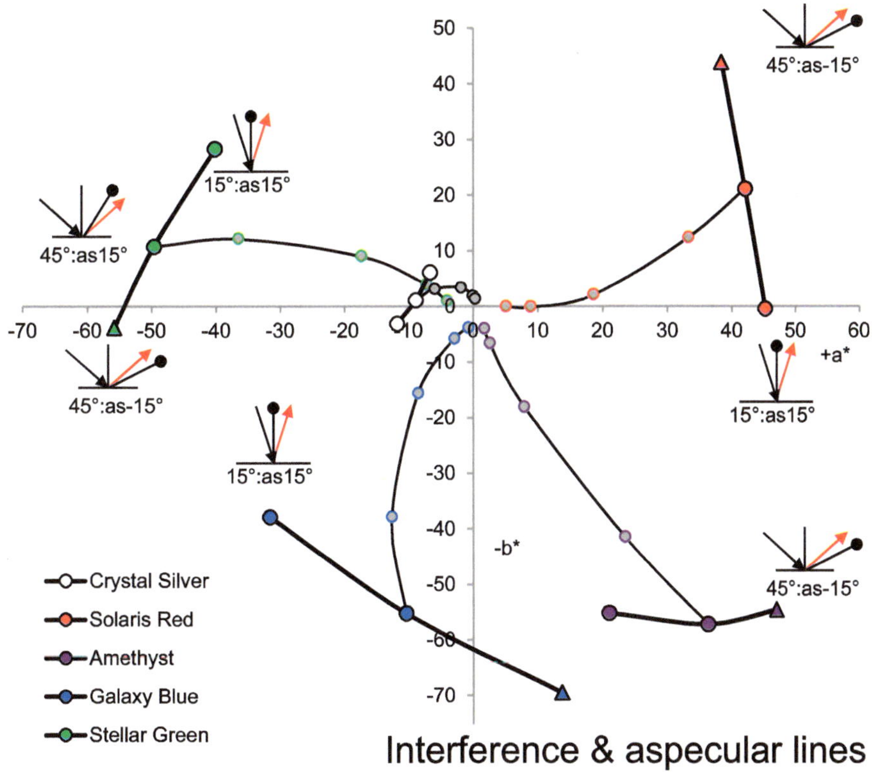

Figure 63: Alumina-based interference pigments with their typical interference and specular lines.

as30°. In Figure 61 and Figure 62, the respective complementary colors yellow-blue and red-green are applied against the specular angles.

The combination of the interference line (15°:as15° – 45°:as15° – 45°:as–15° (= 60°: as15°)) and the aspeular line (45°:as15° to 45°:as110°) serves to identify the interference pigments. The interference line is typical for a pigment (Figure 63).

Interference colors are not pure spectral colors, but always mixed colors.

Orientation

Interference pigments are flat platelets like a sheet of paper. For optimal effect, they should align themselves parallel to the substrate. This is achieved when pigments are used in an automotive base coat, where they can align themselves parallel during application. The pulling method produces uniform results, but due to the direction of pulling, the measurement result depends on the direction of measurement, i. e., the principle of light reversal is not reflected (Figure 64).

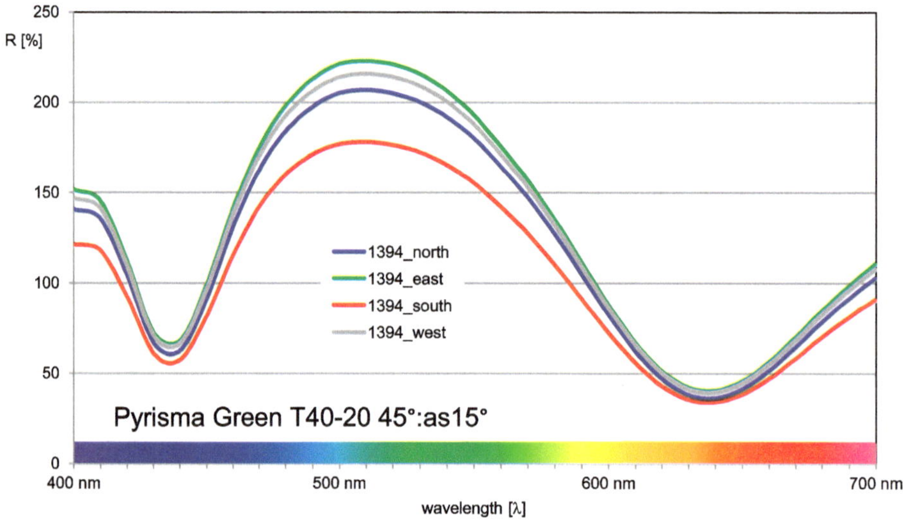

Figure 64: A device can provide different measurement results when it is rotated.

When spraying ChromaFlair 190 – which is representative of all interference pigments – the law of light reversal comes into play. If you swap the illumination and observation, you get almost the same measurements. This turns the flat geometry of 50°:as–65° into a geometry of 65:as–50° and the 5°:as–20° into the steep geometry of 20°:as–5°. The measurement direction is irrelevant here (Figure 65).

If an effect pigment is applied and the measuring device is rotated, the principle of light reversal cannot be applied here due to the orientation of the flakes (Figure 66).

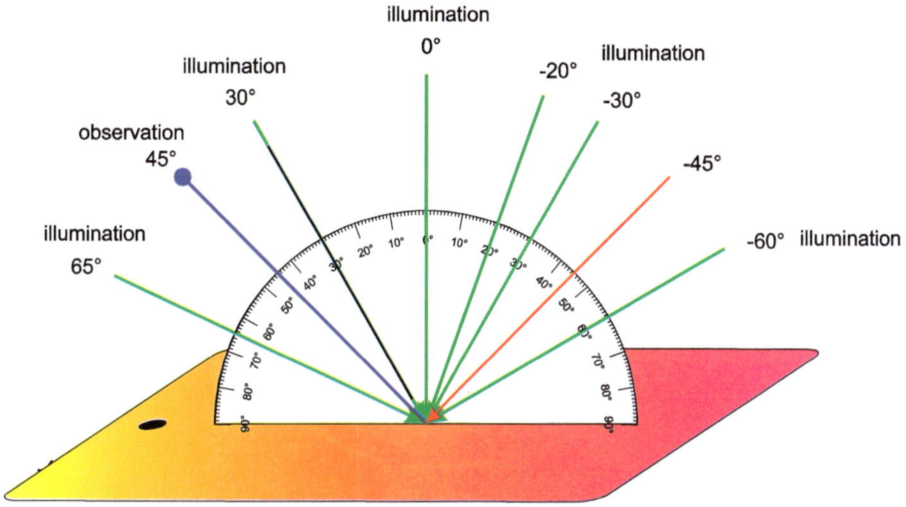

Figure 65: If the angles of illumination and observation are swapped, these geometries result.

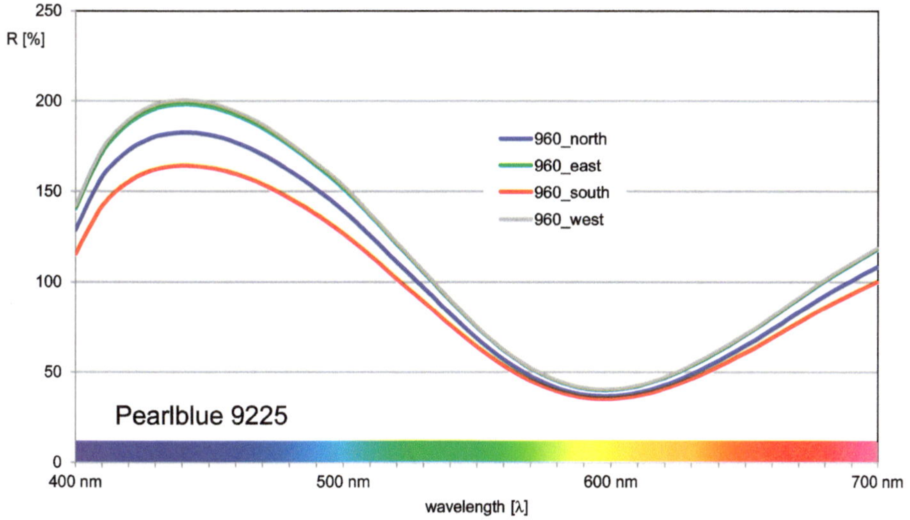

Figure 66: The reflection curves when changing the direction of the measuring device.

The comparison of Figure 67 and Figure 68 reflects the difference: With the sprayed pattern of ChromaFlair190, there is hardly any difference between the measurements at 15° from the specular (cis) and the corresponding illumination angle compared to the measurement results at –15° aspecular (trans). When applying a drawdown with ChromaFlair 190, the differences become clear due to the direction of application.

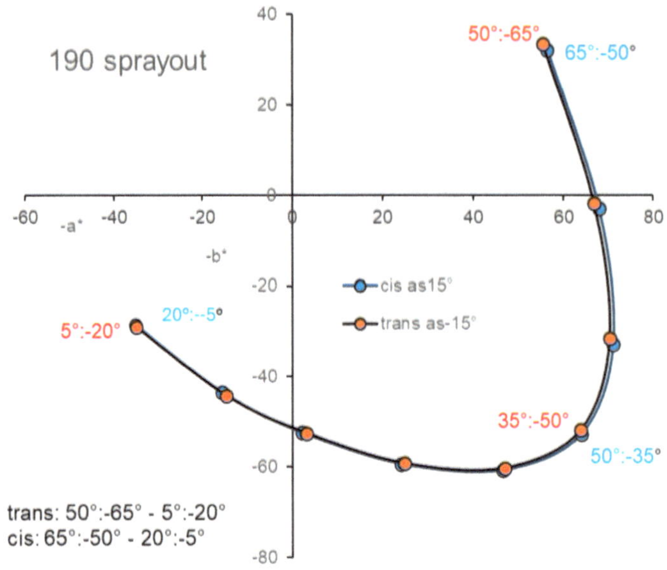

Figure 67: The sprayout of ChromaFlair 190 shows no differences.

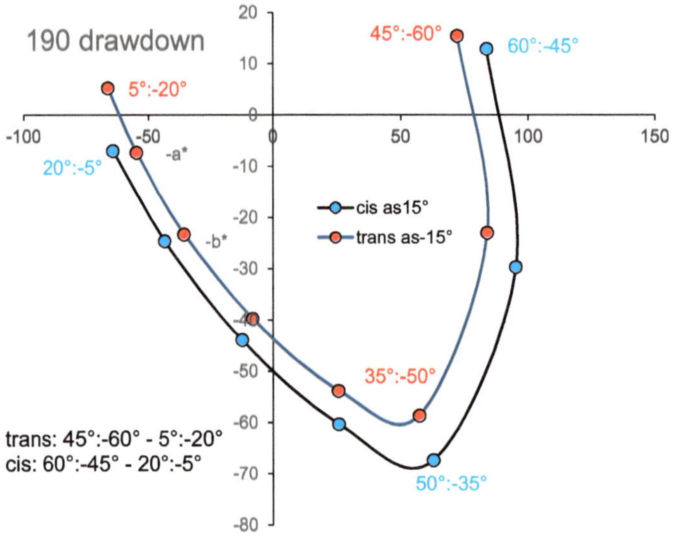

Figure 68: Differences become apparent with drawdown.

In refinishing, so-called mist or effect coats are often applied at the end of the base coat painting process. This involves increasing air pressure or spray speed. This "disorientation" of the flakes influences the effect. Under an electron microscope (Figure 69)

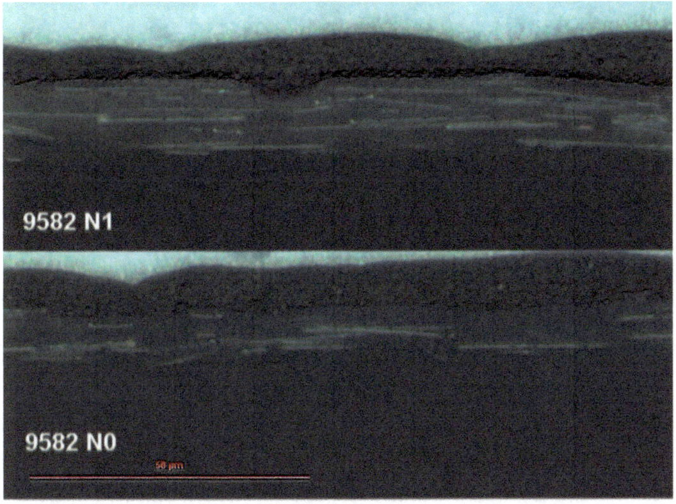

Figure 69: The electron microscope shows the differences when an additional effect layer (N1) is sprayed.

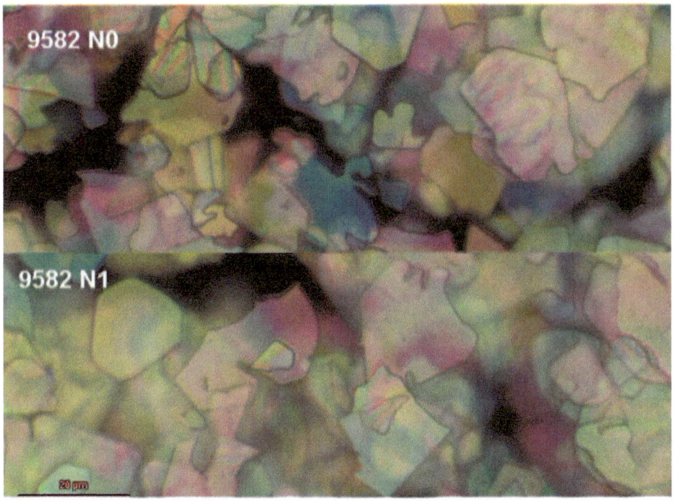

Figure 70: The influence of the effect layer (fog layer N1) can also be seen under a light microscope.

shows the difference in the orientation of the flakes (N0 without effect layer, N1 with effect layer). The difference between N0 and N1 can also be seen under a light microscope (Figure 70).

Size of the flakes

Colored pigments usually have fixed particle sizes. Titanium dioxide is very large, measuring up to 200–400 nm (0.2–0.4 µm), compared to carbon black, which measures 10–100 nm (0.01–0.1 µm). Primary crystals of azo pigments are 20–200 nm (0.02–0.2 µm) in size.

Interference pigments usually have a size range of 10–60 µm. For correspondingly fine versions, this range is 5–25 µm. Interference pigments for cosmetic applications can also range from 20–200 µm.

All values are approximate; for individual pigments, refer to the manufacturer's specifications.

If we compare the measurements of normal interference pigments with their fine versions, the reflection curves of the fine versions are shifted toward shorter wavelengths compared to those of the normal versions (Figure 71, Figure 72). In Figure 73, the brightness values for normal and fine Pearl Blue are compared with the specular geometries. Here, the brightness values for the fine variant are higher than those for the normal variant for all geometries. This is due to the higher scattering effects of the finer flakes. For Pearl Green (Figure 74), the brightness increases in geometries that are far from glossy.

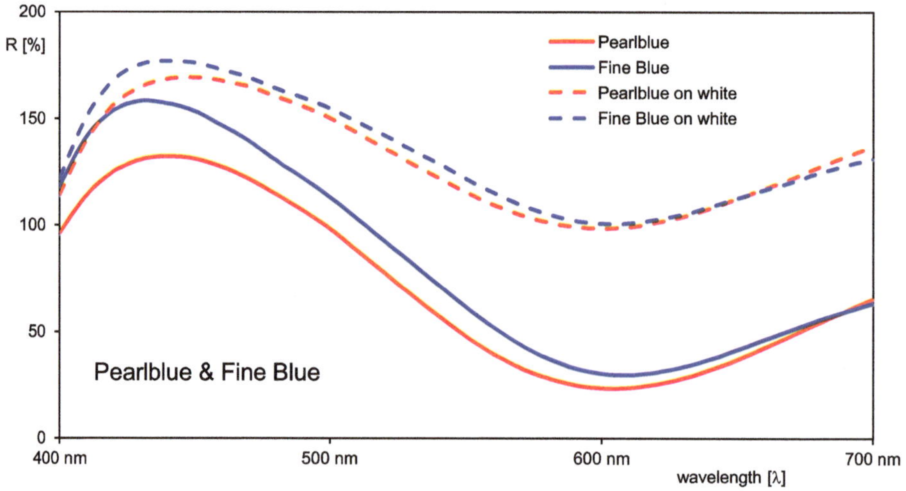

Figure 71: Reflection curves depending on flake size.

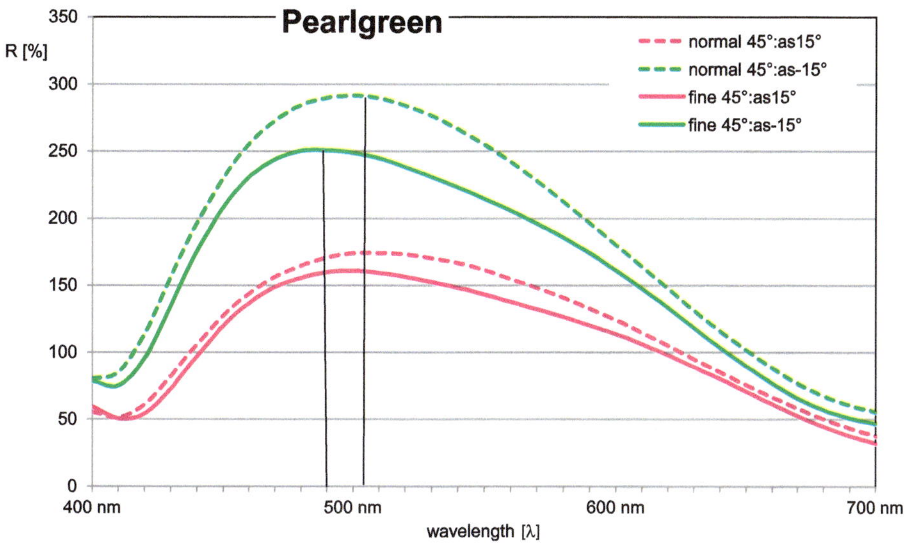

Figure 72: Shift in reflections for finer flakes.

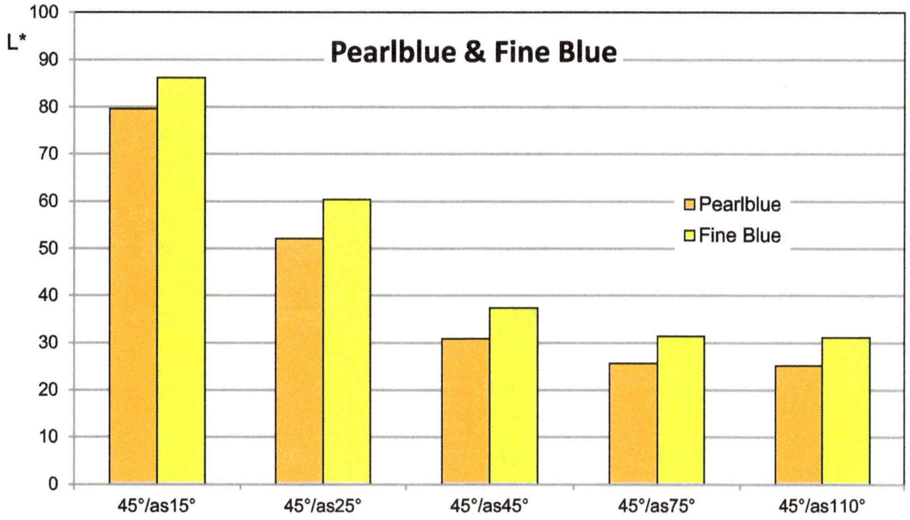

Figure 73: Brightness differences between fine and normal flakes of Pearl Blue 9225.

Concentration

The interference pigments should align themselves parallel to the surface to achieve the optimum effect. When spray-applying in automotive painting, the base coat with the interference pigment is first applied in a two-coat process.

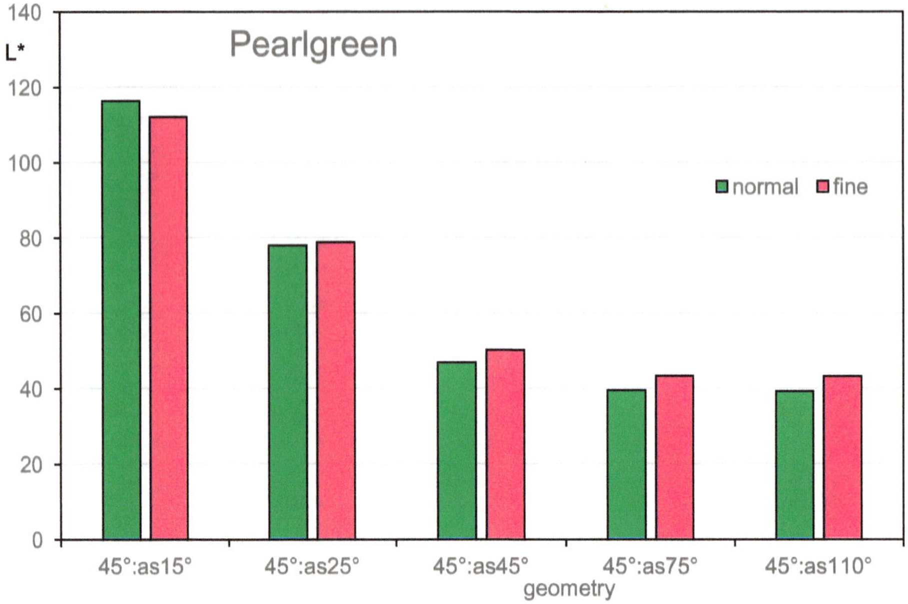

Figure 74: Brightness differences in Pearl Green 9235.

After physical drying through evaporation, the base coat is sealed with clear coat to protect it against chemical and mechanical stress and influences.

The base coat is adjusted and applied in such a way that it remains "open" until the pigment particles are optimally aligned.

In addition to alignment, the question of pigment concentration arises. Ideally, the pigment particles lie next to each other; in the worst case, they lie on top of each other.

In a series of tests, the same Pearl Green was mixed into the paint in various concentrations ranging from 0.25 % to 16 % and applied. The measurements yielded the following results.

The reflections show a clear increase in the maxima close to the gloss. Figure 75 shows the measurement curves resulting from observations at a window or in a light booth. The constant angle of difference between illumination and observation is 30°. The maximum of the reflections increase up to a concentration of 4 %. At higher concentrations, the maximum does not increase further. Only the lateral observation areas with larger aspeular geometries continue to increase, which indicates a "whitening" effect.

The reflection curves at 45°:as15° also show this behavior: they rise to 4 % of Pearl Green, with higher concentrations causing the increase to decrease (Figure 76).

The brightness L* increases with increasing concentration up to 8 % pigment concentration, then remains at the same level at higher concentrations (Figure 77).

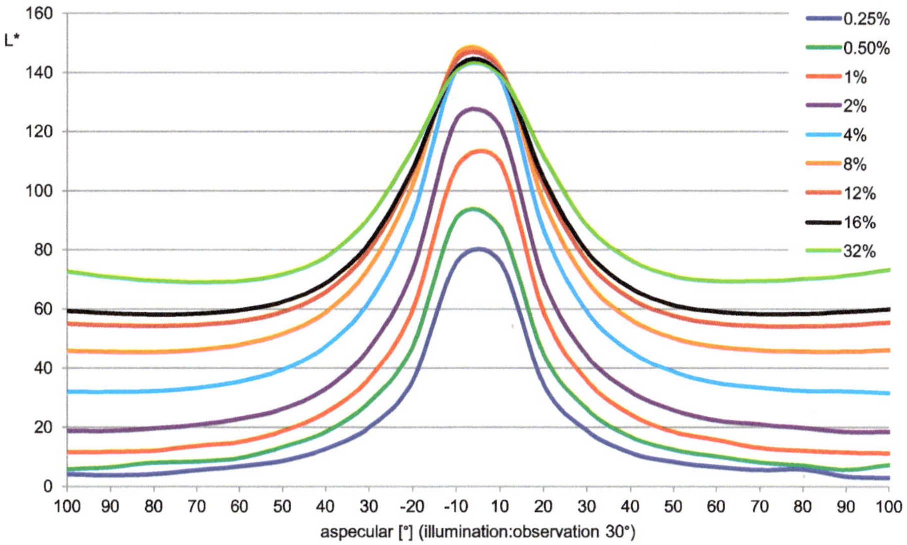

Figure 75: When viewed at the window, the maximum reflection is achieved at 4 %.

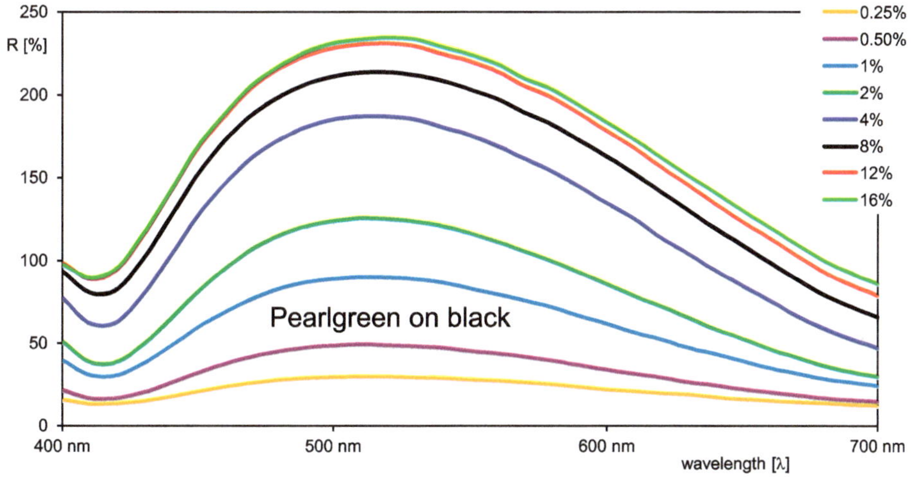

Figure 76: Reflection curves at different concentrations.

The chroma C* increases up to a concentration of 4 % and then decreases (Figure 78).

The a*b* diagram also shows an abrupt color change between 4 % and 8 % pigment concentration (Figure 79).

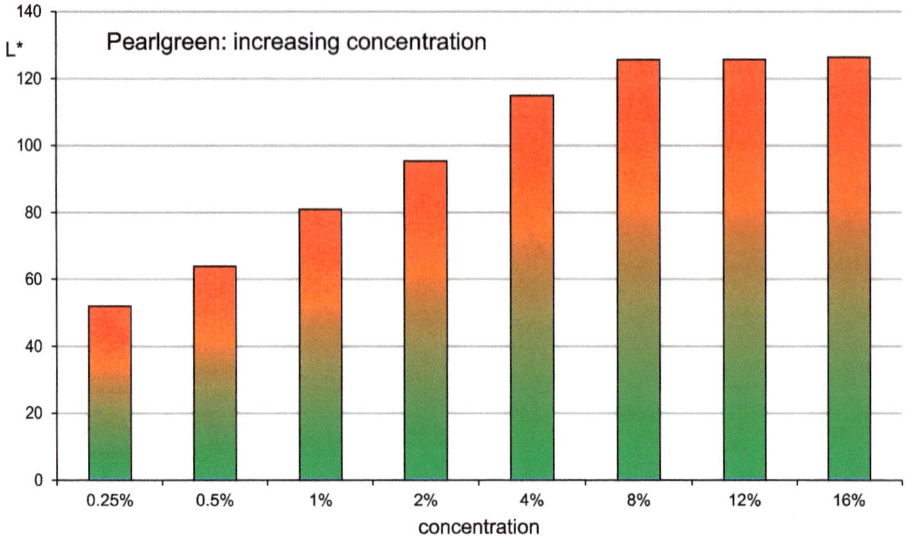

Figure 77: Brightness depends on concentration.

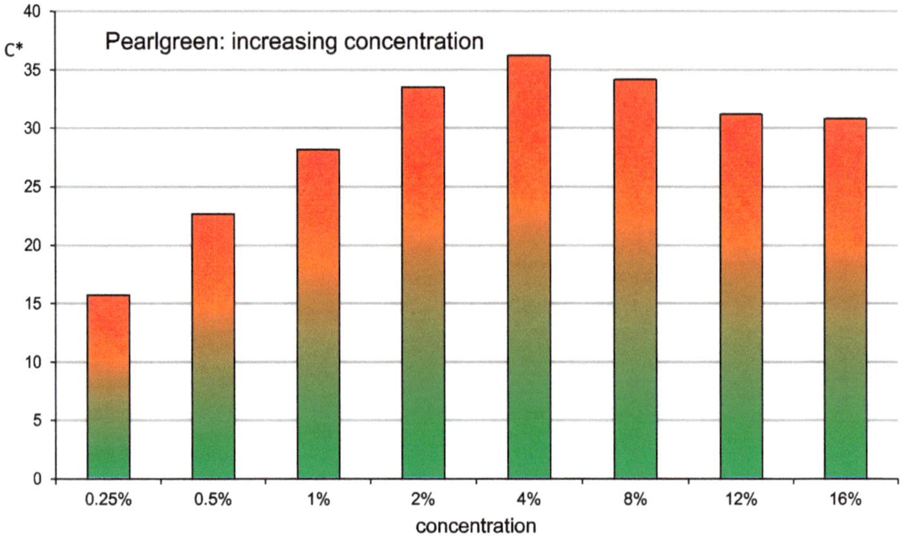

Figure 78: Chroma depending on concentration.

For your own applications, we recommend testing a range of concentrations to determine the optimal concentration.

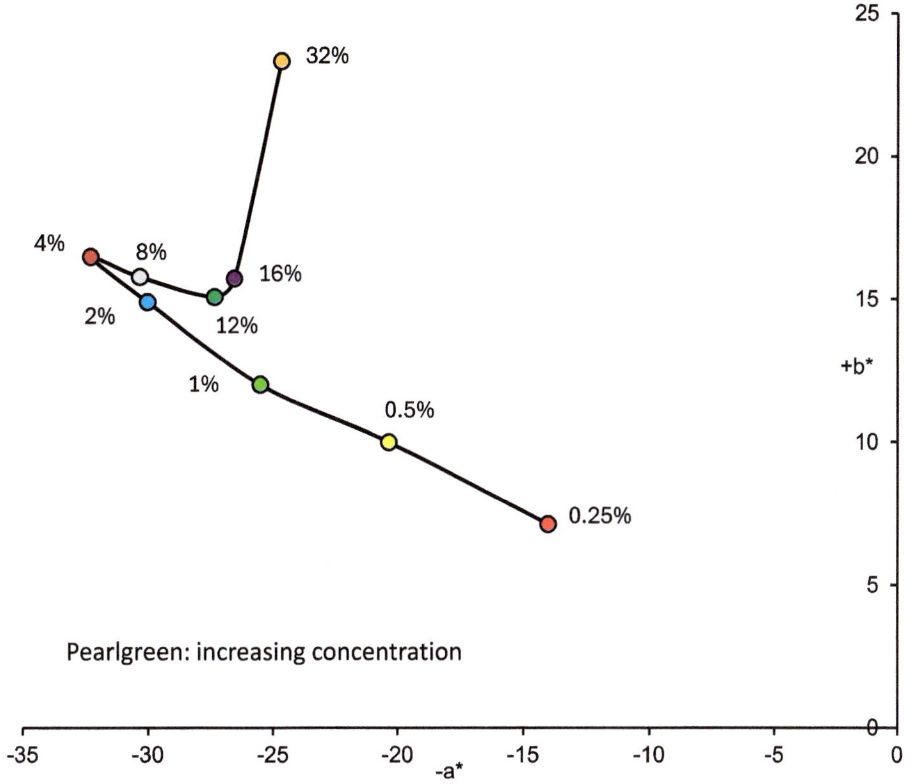

Figure 79: Progression of a*b* values as a function of concentration.

Influence of the substrate color

Due to the transparency of the interference pigments, the resulting color impression also depends on the substrate color. There are fundamental differences between black and white substrates, as reflected by the complementary transmission color. This can be observed and measured. On a black substrate, the reflection colors are similar to those on a white substrate.

Experiments with pastel-colored substrates and the same base coat result in significant shifts in reflection. This applies to all interference pigments.

The base color also plays a role in refinish or OEM paints if they are transparent. In the example of the Peugeot EKQ (Lucifer Red) series color, three colors from the mixing system were mixed with white in the ratios 80:20, 60:40, 40:60, and 20:80 (Figure 80). In the a*b* chart, the typical progression with a turning point is particularly evident when green is mixed with white.

These blends with white and the original mixed colors were sprayed onto metal panels as base colors. Then the same base coat with the series color Peugeot EKQ was

Figure 80: White blends as base colors.

Figure 81: Peugeot EKQ base coat was sprayed over all colored substrates.

applied to all blends at the same time and then sealed with clear coat (Figure 81). As a comparison sample, a panel was sprayed with the base coat until the color was opaque (using a black and white control marker). The red series color shows color shifts when the substrate color does not exactly match the series color. The transparent red series color acts like a red filter. It mainly allows red light to pass through, which is reflected by the substrate. However, the red substrate does not reflect the red light in the same way as the series color, resulting in color shifts (Figure 82). This behavior is clear from the re-

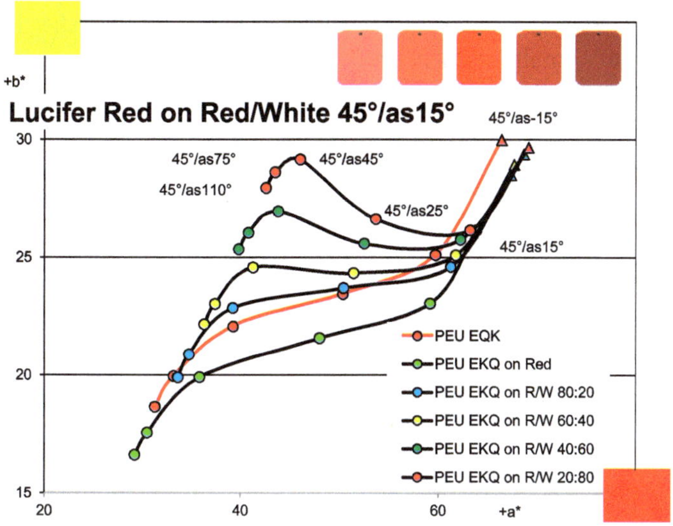

Figure 82: Lucifer Red on red:white.

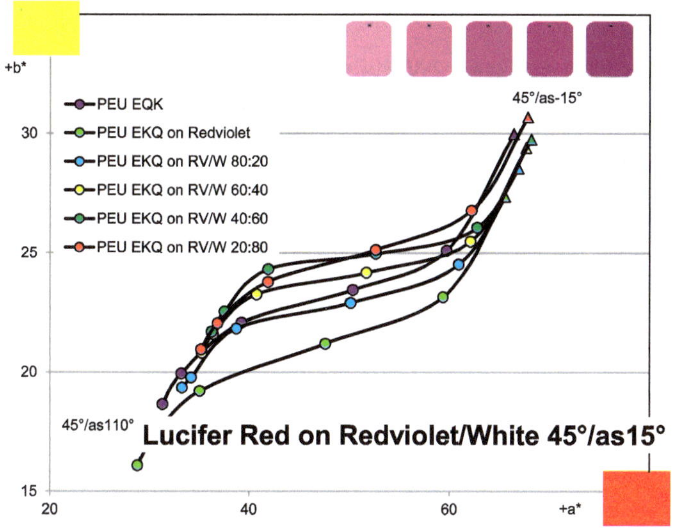

Figure 83: Lucifer Red on violet:white.

flection curves and the a*b* charts. Significant differences are particularly apparent in the nonglossy geometries. The influence of the red-violet substrates is less pronounced (Figure 83). Both the a*b* color values and the reflections are much closer to the original in terms of color.

The results for green substrates are very interesting. Since the red light is reflected unchanged by the base coat – similar to a neutral color – there is almost no color shift in the series color. This applies to the largest and smallest white blends of green (Green:White 80:20 and 20:80) (Figure 84).

The fact that the results on gray substrates are not as good as those on green substrates is due to the different gray mixtures. With all grays, the reflections decrease in

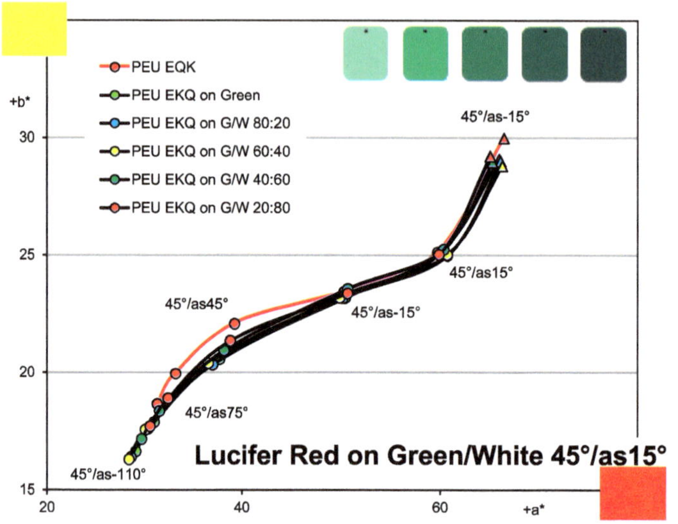

Figure 84: Lucifer Red on green:white.

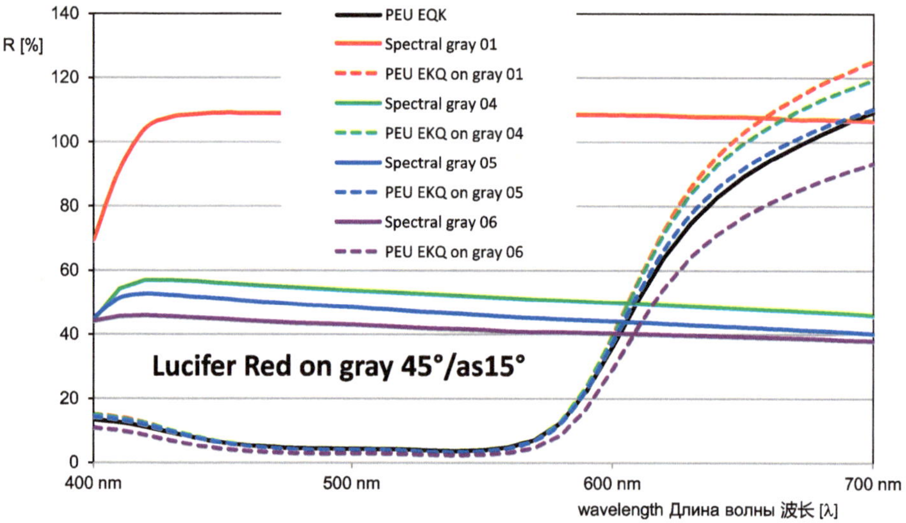

Figure 85: Reflection curves on gray.

the longer wavelength spectral range, so that differences occur in the reflection of the red light of the colored base coat (Figure 85).

In a series of tests, base coats with interference pigments were sprayed over pastel-colored substrates. Here, the influence of the substrate color on the overall color impression is clearly visible.

Portfolio

Interference pigments can be roughly distinguished according to their manufacturing process: In the wet chemical process, natural or synthetic mica is commonly used as a carrier material (Iriodin, Afflair, Lumina, KC Pearls, Syakarp, Plovence, Setallic, Kyntaline, XillaMaya, ChesPearl, ChesCrystal, etc.), which is coated with titanium dioxide or iron oxide. Depending on the duration of the process, the layer thickness of the metal oxide increases, which influences the resulting color.

The layer thickness of the applied material determines the resulting reflection color. As the layer thickness increases, titanium dioxide changes from white to yellow and then to red, violet, blue, and finally green. Some manufacturers distinguish between white pearls and interference pigments, which is a marketing strategy and has nothing to do with the structure of the interference pigments.

Other carrier materials include aluminum oxide (Xirallic), silicon dioxide (Colorstream), and borosilicate (Luxan, Borostar). They are also mainly coated with titanium dioxide or iron oxide, as well as combinations of both metal oxides. Coatings with iron oxide result in red and reddish-brown pigments, while coatings with combinations of titanium dioxide and iron oxide result in golden pigments. A sample coating was applied using a pigment formulation based on the combination pigments. The result was a combination of reddish to light gold colors that looked deceptively similar to real gold. This “aurum magicum” was significantly “goldener” than comparable colors. Due to its play of colors, most viewers considered it to be real gold (Figure 86).

Figure 86: “aurum magicum”—magical gold based on interference pigments.

Nontransparent pigments based on aluminum are produced as carrier materials in a high vacuum. In the case of ChromaFlair pigments, for example, a highly refractive layer is first applied to transparent film. This is followed by a second layer and then a third layer of aluminum. The fourth layer corresponds to the second and the fifth to the first. The film is then crushed and the plastic material is washed out.

Other examples are Paliocrom and Zenexco pigments. They exhibit similar optical properties to transparent interference pigments but lack transmission color. Their color also shifts toward shorter wavelengths when illuminated more flatly.

The names used for classic mica pigments in particular have become standard among many manufacturers. Some of these are still based on the old names used by DuPont. The prefix "9" stands for exterior pigments. The following "1" indicates the different fraction sizes of white interference pigments. A subsequent "2" indicates colorful interference pigments. The other numbers stand for the corresponding color.

Table 1: Colors produced by mica pigments with different coatings (layer).

Number	Color	Layer
9103	Silver white	TiO_2
9121	Silver white	TiO_2
9205	gold (yellow)	TiO_2
9215	red	TiO_2
9217	lilac	TiO_2
9225	blue	TiO_2
9235	green	TiO_2
9300	gold	TiO_2/Fe_2O_3
9302	fine gold	TiO_2/Fe_2O_3
9500	red brown	Fe_2O_3
9507	Scarab red	Fe_2O_3

Special interference pigments include liquid crystals and interference pigments with a ribbed structure. Liquid crystals are chemical compounds that, in addition to their solid and liquid states of aggregation, also have a mesophase. In this phase, they behave optically like crystals with birefringence and polarization but are liquid.

Physicist Otto Lehmann (1855–1922) studied them extensively and published his work "Flüssige Krystalle" (Liquid Crystals) in 1904. He expanded his microscope with a heatable plate to research the liquid-crystalline phases. The microscope and many of his samples have been preserved and can be admired in the Merck Museum (Figure 87).

To demonstrate how colors depend on temperature, I measured the colors of black cardboard boxes that I sprayed with liquid crystals. I placed these boxes on a measuring device (with the measuring opening at the top) and weighed them down with a glass container (aquarium). I filled the glass container with hot water and measured

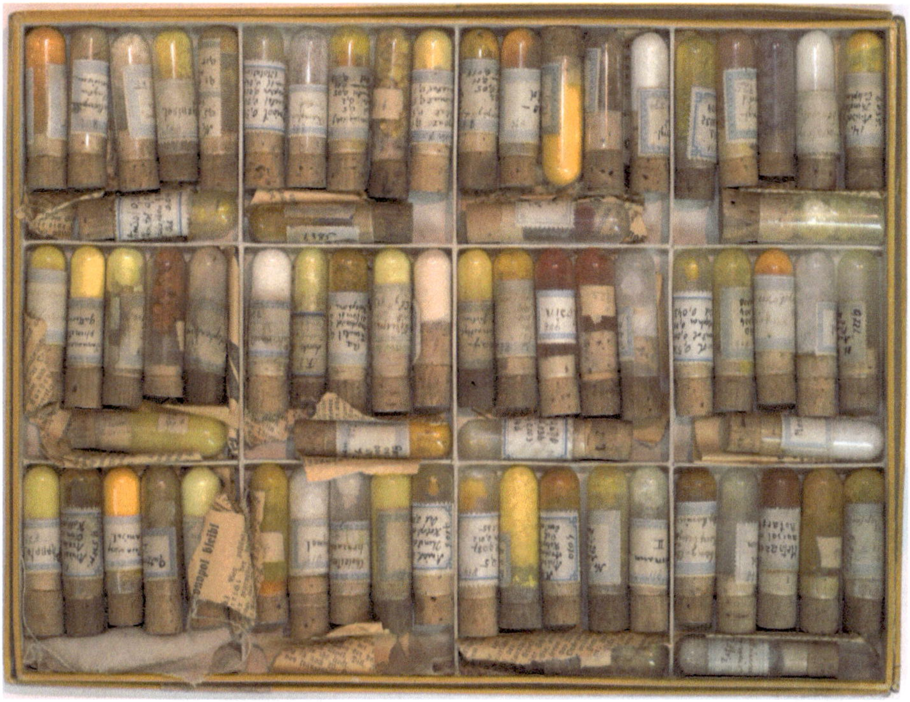

Figure 87: Samples with liquid crystals by Otto Lehmann.

its temperature continuously. The temperature of the water and the cardboard underneath decreased in 1 °C increments. At the same time, color measurements were taken using different geometries (illumination/observation). The resulting diagrams impressively show the color gradient as a function of temperature: Figure 88 shows the color change in relation to the temperature change from 44 °C to 26 °C at illumination angles of 65° and 45°. At 29 °C, the color shifts when illuminated at a flatter angle. The reflection curves shift toward shorter wavelengths as the temperature rises (Figure 89).

Liquid crystals exist as nematic phases (light-dark), smectic phases, or cholesteric phases. The former were initially used to display the time in digital watches. The latter are used to display the temperature in aquariums or in medicine. In polymers, cholesteric phases are no longer temperature-dependent (nonthermotropic). They are found in almost all PC, laptop, and smartphone screens.

SpectraFlair diffractive pigments are a special type of interference pigment. They have a structure like a washboard (Figure 90). Multiple refraction causes the light to be split into a narrow angle range.

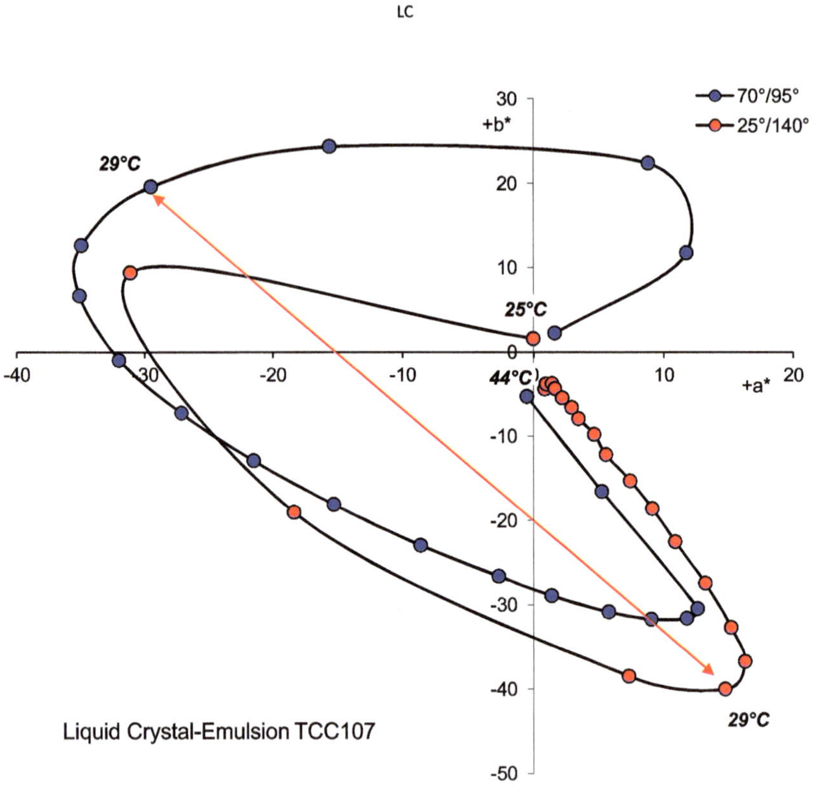

Figure 88: Color changes depending on temperature.

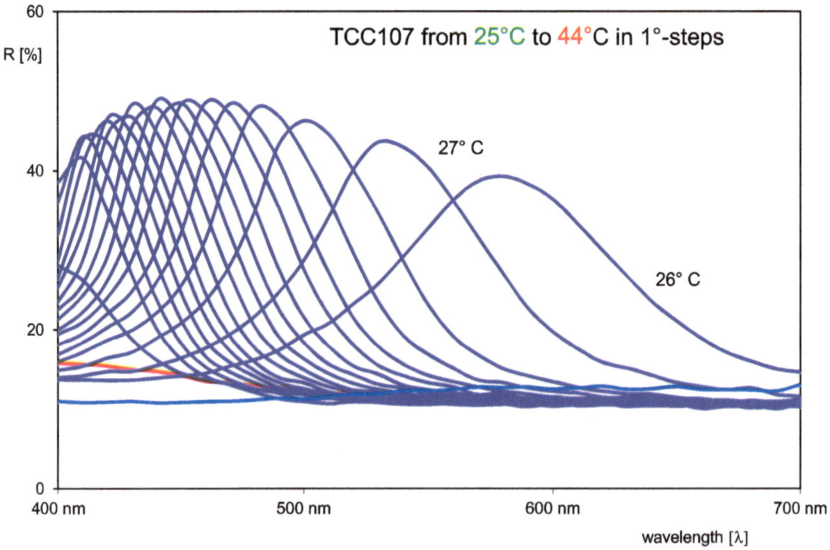

Figure 89: Reflection shift to shorter wavelengths with increasing temperature.

Figure 90: Washing board structure of the SpectraFlair.

4 Mixing behavior

One of the main tasks of a paint laboratory is to mix pigments to create the desired color or reproduce a specified color. Two mixing behaviors come into play here: colored pigments mix subtractively (actually multiplicatively) and interference pigments mix additively. If all colored pigments were mixed, the result would be a very dark gray. If the same were done with interference pigments, the result would be white.

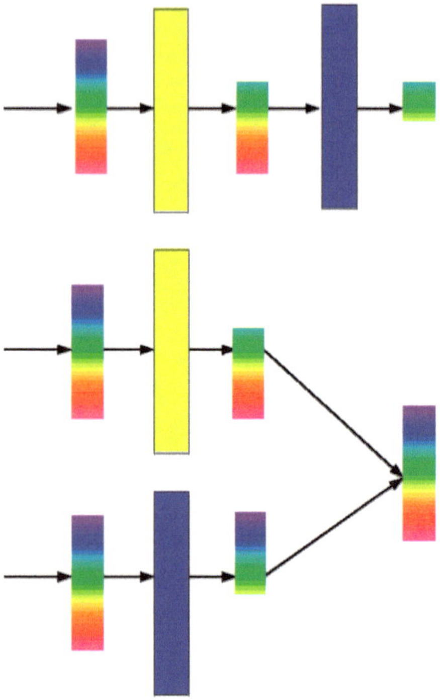

Figure 91: Subtractive (top) and additive mixing.

To understand both mixing behaviors, the light paths in Figure 91 is shown. If white light is sent through a yellow filter and then a blue filter in succession (above), the result is green, which is transmitted by both filters (subtractive mixing). Both the blue and the yellow contain green, which remains "left over" when mixed. If the filters are placed in parallel and white light is sent through them (below), the respective proportions that are transmitted by each of the filters combine again to form white (additive mixing).

When mixed subtractively with colored pigments, yellow and blue produce green, while yellow and blue interference pigments mix to produce white.

Interference pigments are often used in blends with other interference pigments, with colored pigments, or with aluminum pigments. There are a few properties to keep in mind here.

https://doi.org/10.1515/9783111439204-004

Blends with interference pigments

Interference pigments are colorful individualists. They have their own color and special effect, which usually makes them irreplaceable. Their color can be influenced by mixing similar interference pigments, as these examples show:

9205 + 9225, 9215 + 9235

Since interference pigments with titanium dioxide are available as coating materials in all colors of the color wheel, colors can also be mixed between them. They not only mix additively, but also almost ideally. For example, while a 1:1 mixture of a yellow color pigment with a red color pigment does not produce orange, which is halfway between the two colors, a 1:1 mixture of a yellow and a red interference pigment is almost in the middle and equidistant between the two. Figure 92 shows the color circles ("kites") of interference pigments based on mica and alumina coated with titanium dioxide.

A 1:1 mixture of interference yellow and interference blue yields white, in contrast to the mixture of yellow and blue color pigments, which yields green (Figure 93, Figure 94). Interference red and interference blue mix to form violet, while the subtractive mixture of red and blue yields a very dark brown (Figure 95, Figure 96).

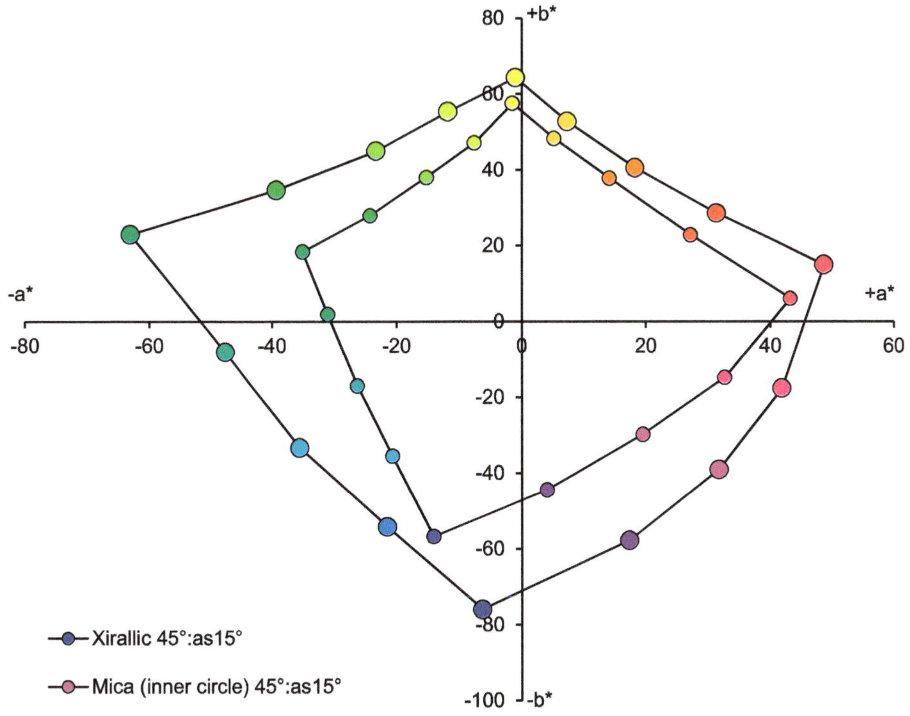

Figure 92: Interference pigments mix additively.

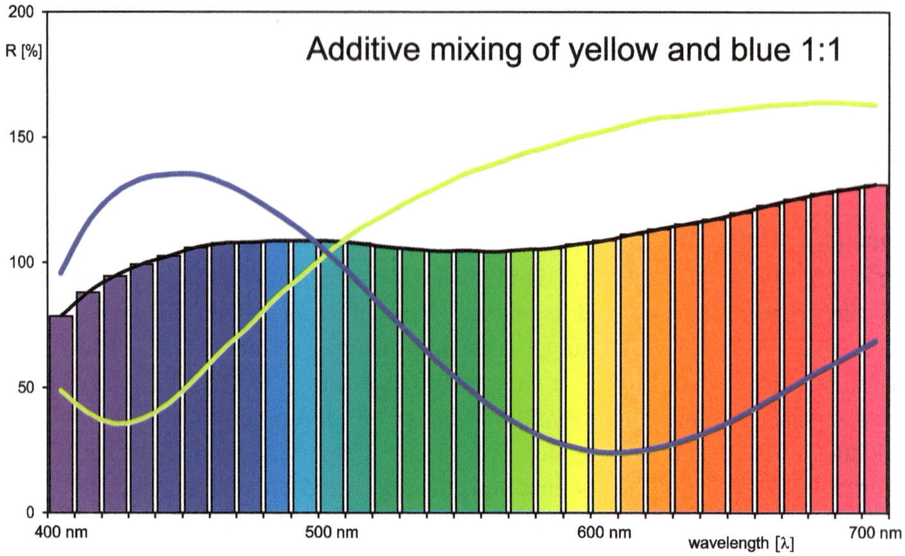

Figure 93: Additive mixture of interference yellow and blue.

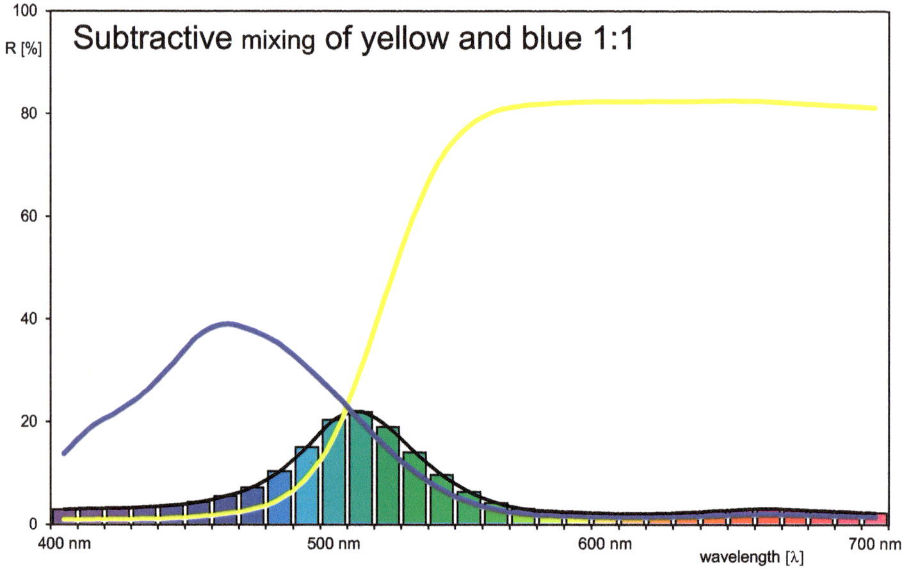

Figure 94: Subtractive mixing of yellow and blue.

The 1:1 mixture of interference green and interference red produces yellow. In both cases, the direction of the interference line rotates 180° when the interference green becomes an interference red: The interference green changes from yellowish to bluish

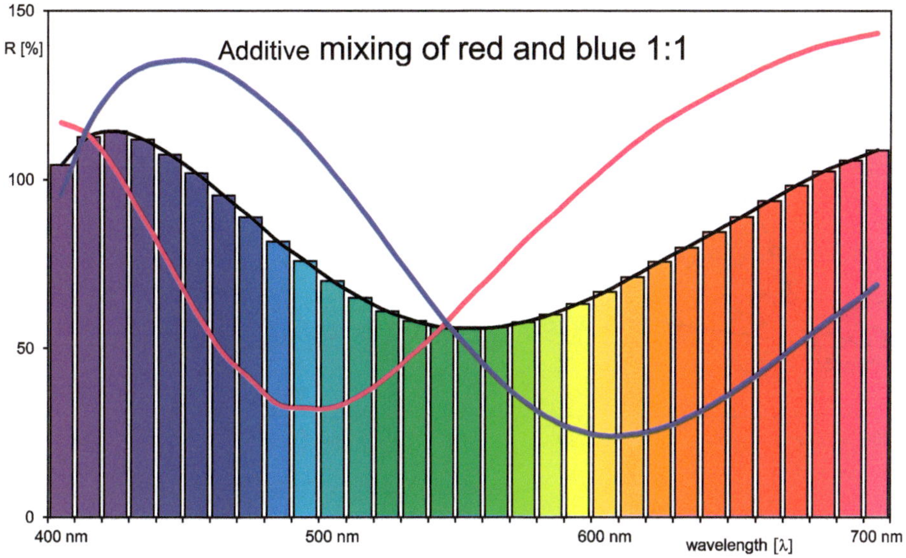

Figure 95: Additive mixture of red and blue.

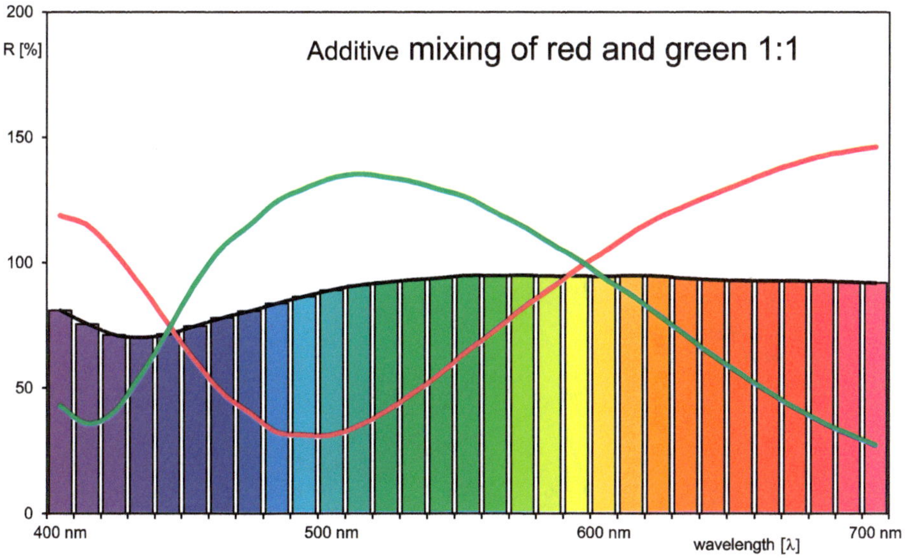

Figure 96: Additive mixture of red and green.

when illuminated more flatly. Correspondingly, interference red changes from bluish to yellowish (Figure 97).

A yellow interference pigment changes color from reddish to greenish when illuminated at a flatter angle. Similarly, a blue interference pigment changes from greenish

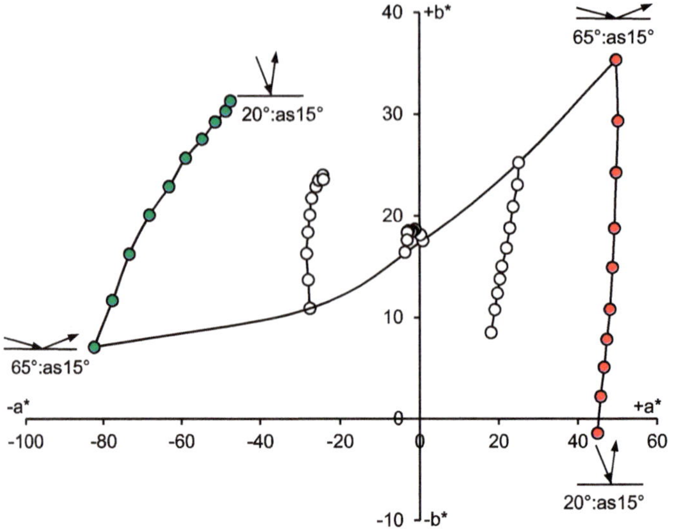

Figure 97: When interference green and interference red are mixed, the interference lines “rotate.”

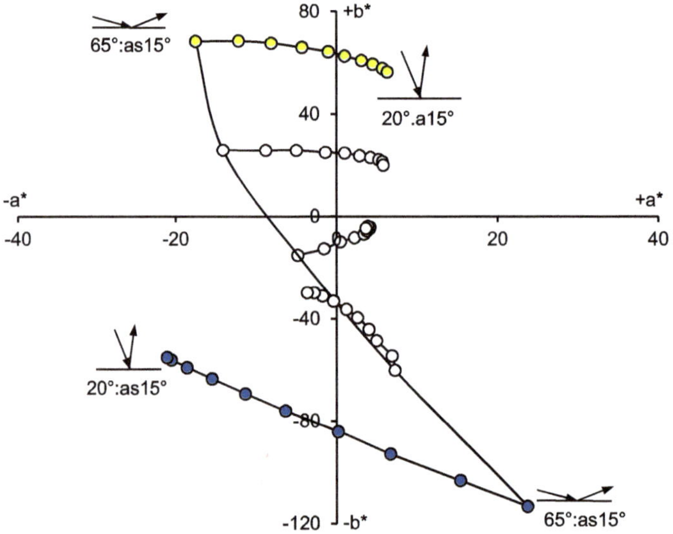

Figure 98: When mixing interference yellow and blue, the interference lines also “rotate.”

to reddish. When mixed, the interference line rotates by almost 180° (Figure 98). The reflection curves in Figure 99 reflect the mixing behavior: The initial colors form two nodes through which the mixture also passes.

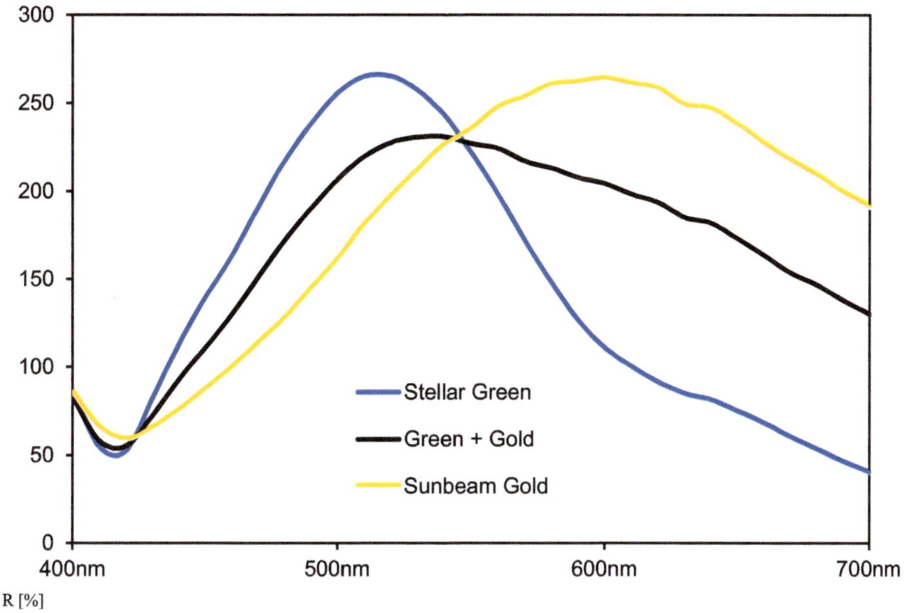

Figure 99: Reflection curves of interference pigments intersect at nodes through which their blends also pass.

In the mixtures, the respective interference line is shortened. The corresponding aspecular lines are also shortened. At a transition point, the mixture has almost no interference line and no aspecular line.

This change can be illustrated by plotting the aspecular values of a* or b* against the aspecular angles. The red-green change is described by the a* values and their change, while the yellow-blue change is described by their b* values (Figure 100, Figure 101).

The Viola Fantasy exhibits similar behavior, consisting of a green-yellow section ranging from flat to classic and a yellow-violet section ranging from classic to steep illumination. Through additive mixing, the green-yellow section is canceled out when the Viola Fantasy is mixed with the violet 9219 (Figure 102). This shifts the interference line to the violet part of the a*b* diagram. “Conversely,” the yellow-violet part can be achieved in the green by adding green 9235 through additive mixing (Figure 103).

Yellow lies between the two “end colors” of Viola Fantasy Green and Violet. This can also be “removed” by adding blue 9225 (Figure 104). Here, the interference line of Viola Fantasy shifts toward the achromatic zero point (= white). Corresponding color reactions occur with yellow 9205 and red 9215.

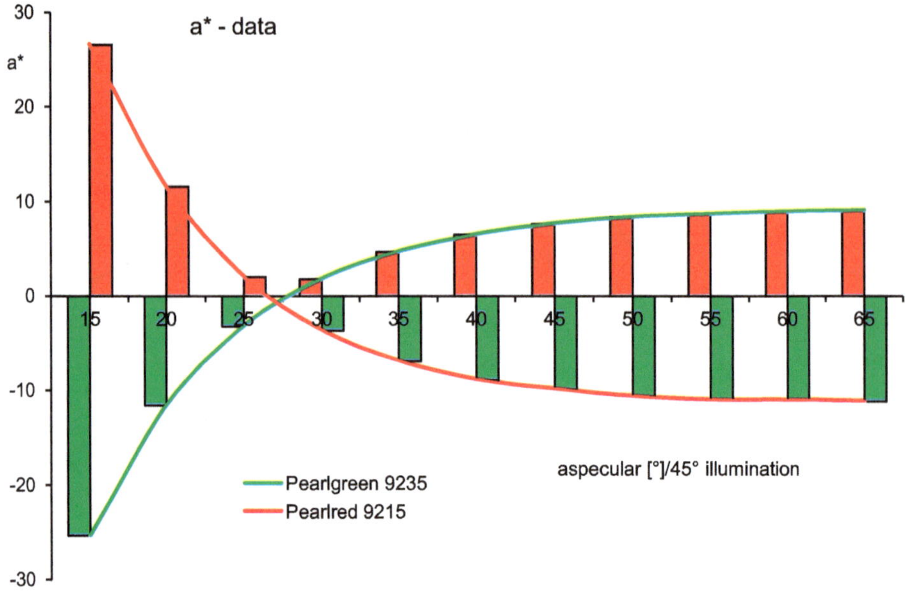

Figure 100: Interference green and red colors change to their complementary colors.

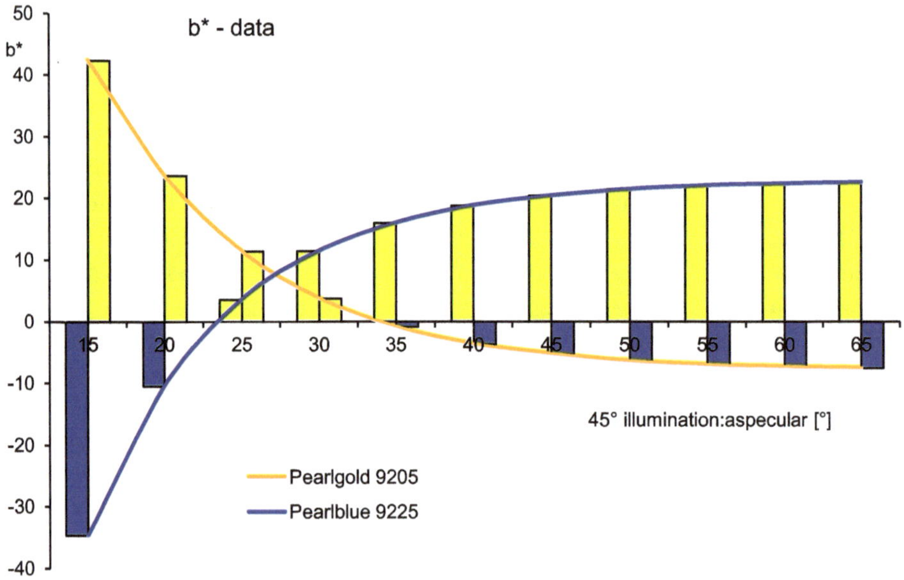

Figure 101: Interference yellow and blue colors change to their complementary colors.

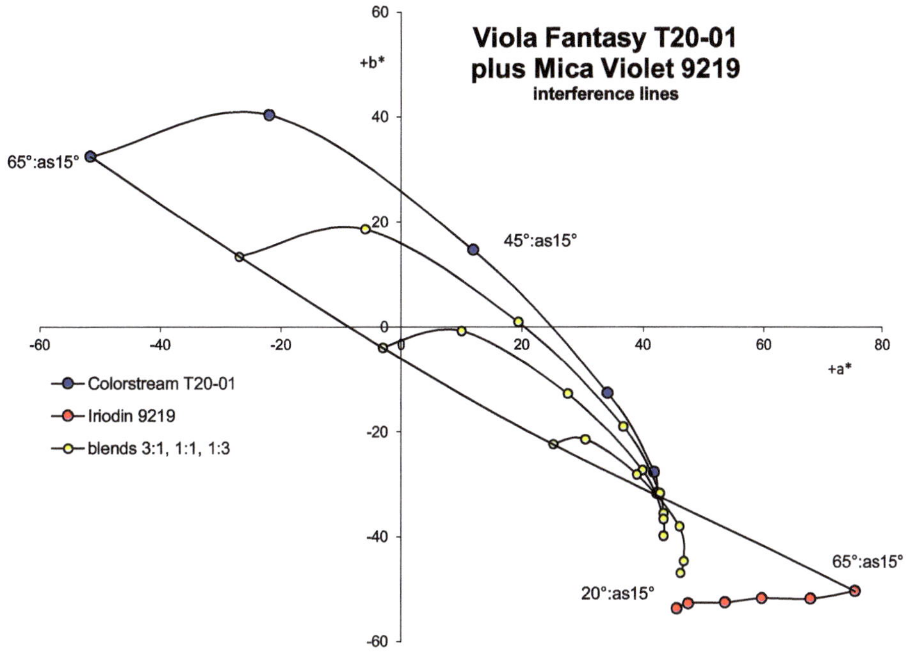

Figure 102: The colors of Viola Fantasy shift to violet with Pearl Violet.

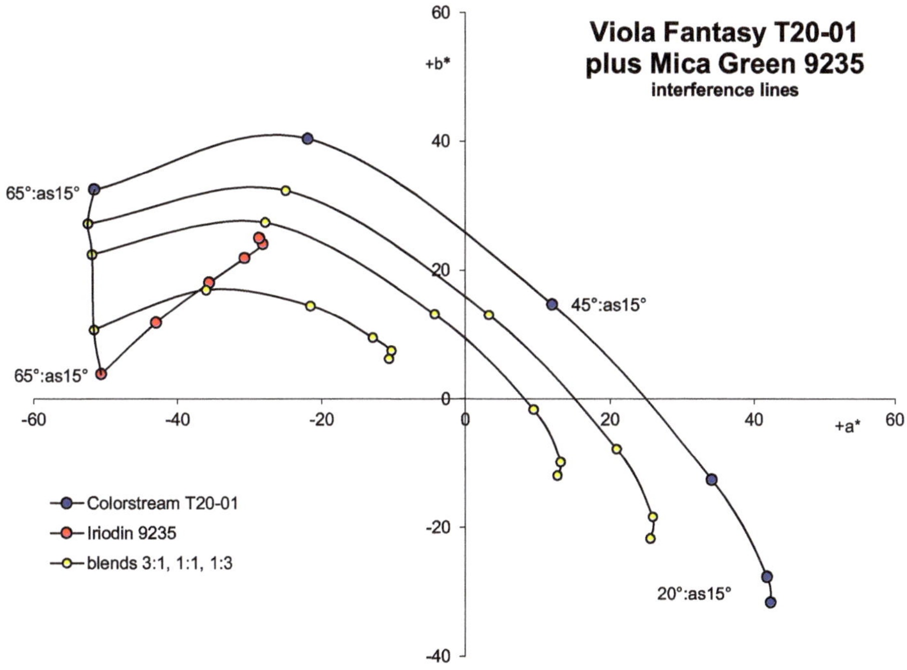

Figure 103: The colors of Viola Fantasy shift to green with Pearl Green.

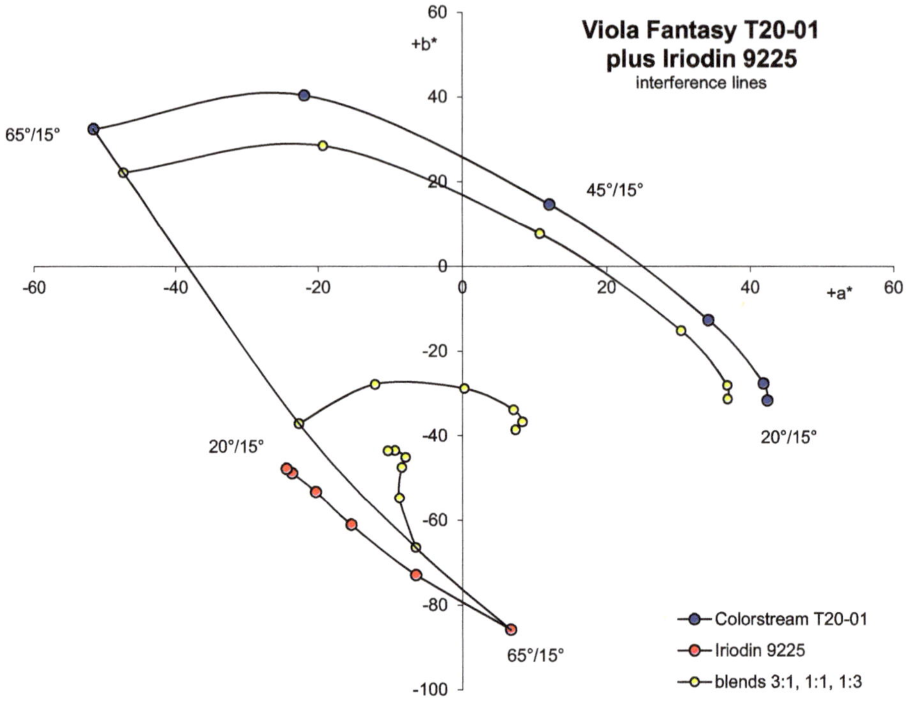

Figure 104: The colors of Viola Fantasy shift toward achromatic with Micablue.

9205 + 9215 + 9225 + 9235

When interference pigments are mixed with titanium dioxide coatings in ratios of 3:1, 1:1, and 1:3, a color wheel is created. Mixing behavior can be exploited even further. Starting with one interference pigment, it is mixed with the opposite 2-pigment mixtures of interference pigments (Figure 105). Or you can also mix the pairs of mixtures with opposite pairs of mixtures. In any case, the ideal mixing behavior of the interference pigments becomes apparent. This goes so far that a mixture of yellow, blue, and red interference pigments produces the same color as a mixture of yellow, green, and red interference pigments. On a white substrate, the mixtures change to their complementary colors (Figure 106), the aspecular lines run toward achromatic colors on a black substrate (Figure 107).

The reflection curves help to understand the mixing results: The reflection curves of the initial pigments intersect at one or two nodes. All mixtures "oscillate" through these points (Figure 108, Figure 109).

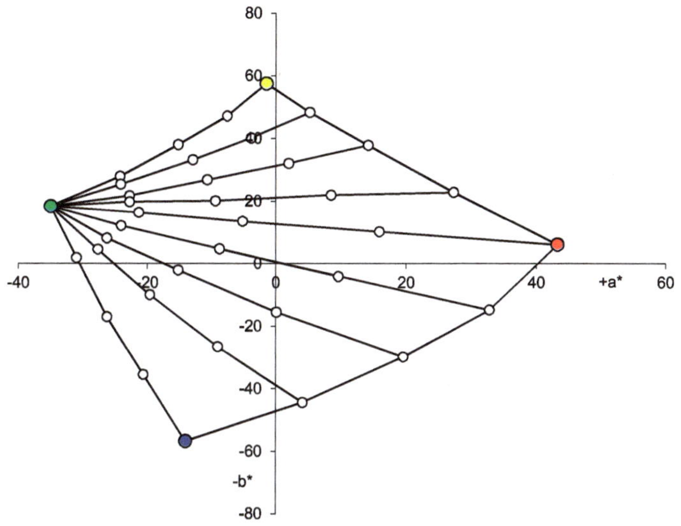

Figure 105: Mixtures of an interference pigment with opposite mixtures in the color wheel.

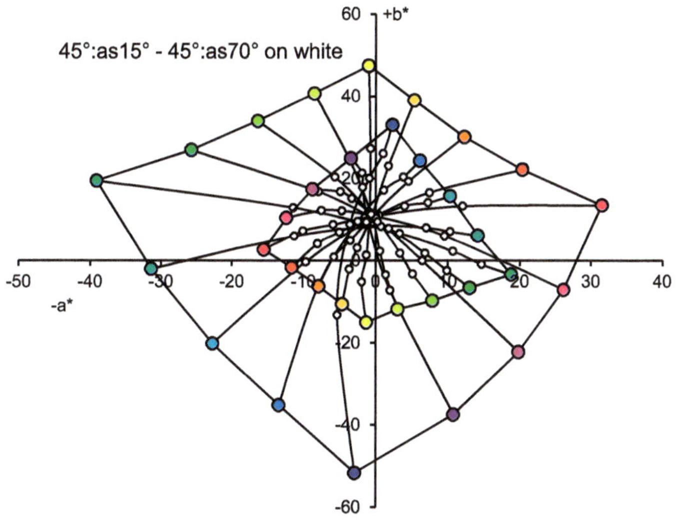

Figure 106: Aspecular lines of interference pigments and their mixtures on a white substrate run from the reflection color to the transmission color.

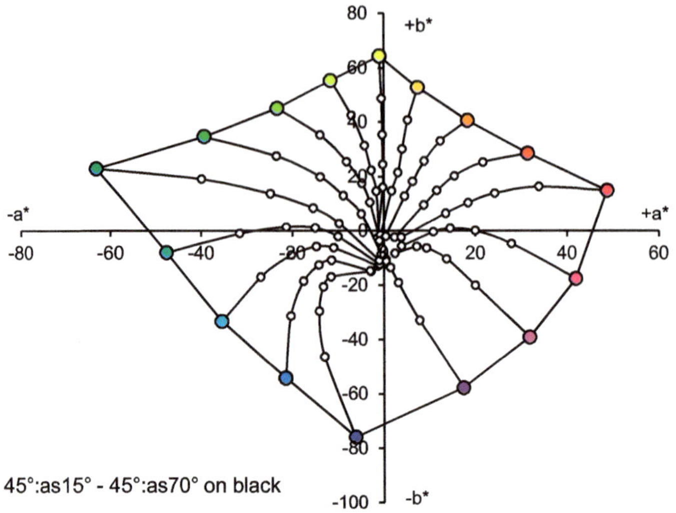

Figure 107: On black substrate, the aspecular lines of the same pigments run toward achromaticity.

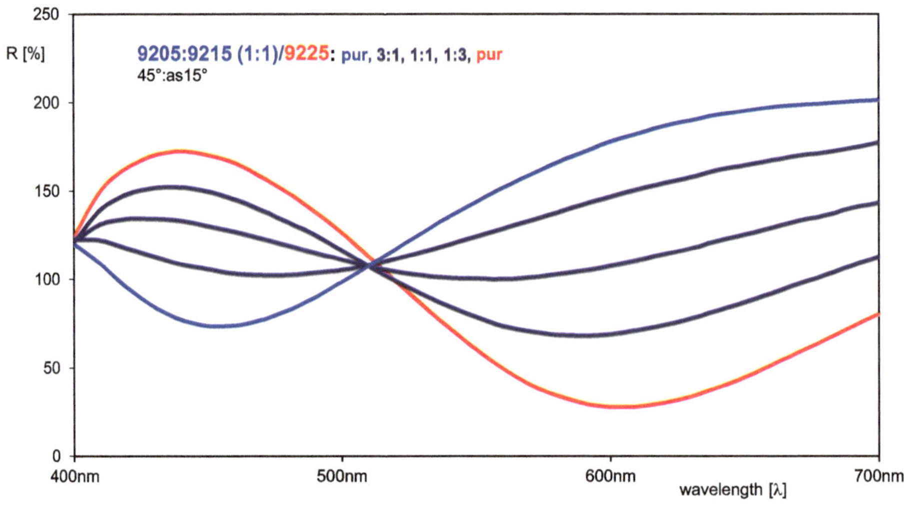

Figure 108: Reflection curves of ternary mixtures pass through nodes.

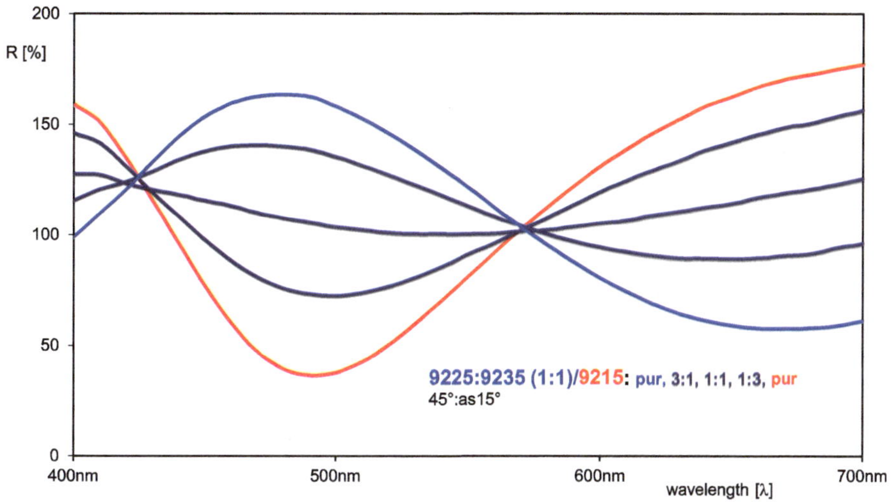

Figure 109: Mixtures of interference red, blue, and green behave ideally.

9507 + 9205

Blends of Scarab Red 9507 with Interference Yellow 9205 produce an orange close to specular (Figure 110). The 9507 is close to specular green that blends with yellow to form orange. The explanation can be found in the reflection curves: The 9507 has a clear minimum in the yellow spectral range, which is gradually filled in by the 9205 (Figure 111).

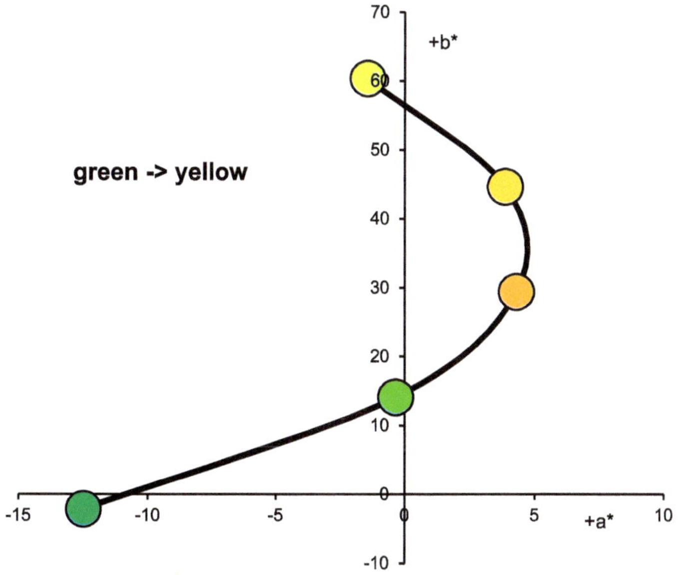

Figure 110: The scarab pigment blends with interference yellow to form orange, close to gloss.

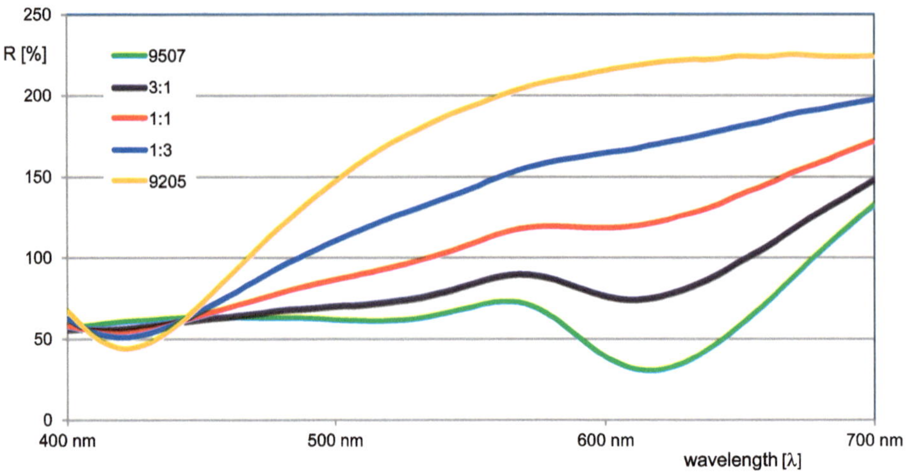

Figure 111: The reflection curves show the "filling" of the mixtures by the interference yellow.

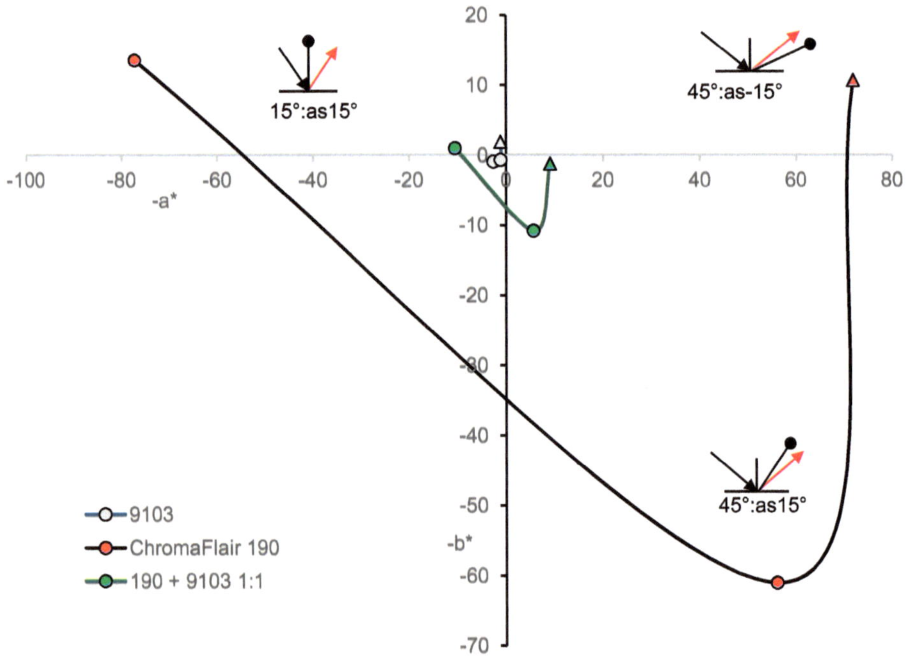

Figure 112: ChromaFlair blended with Pearl White.

Interference can be shifted in color, as demonstrated by the two interference pigments ChromaFlair 190 and Colorstream Viola Fantasy. ChromaFlair consists of a long arm of the interference line from steep to classic illumination from green to violet. The

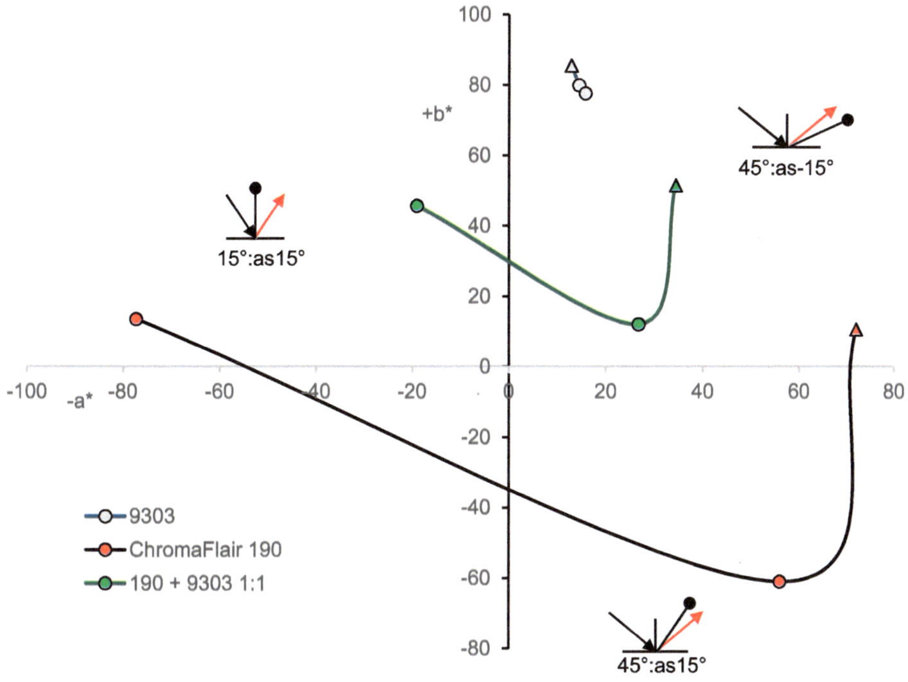

Figure 113: ChromaFlair blended with Gold 9303.

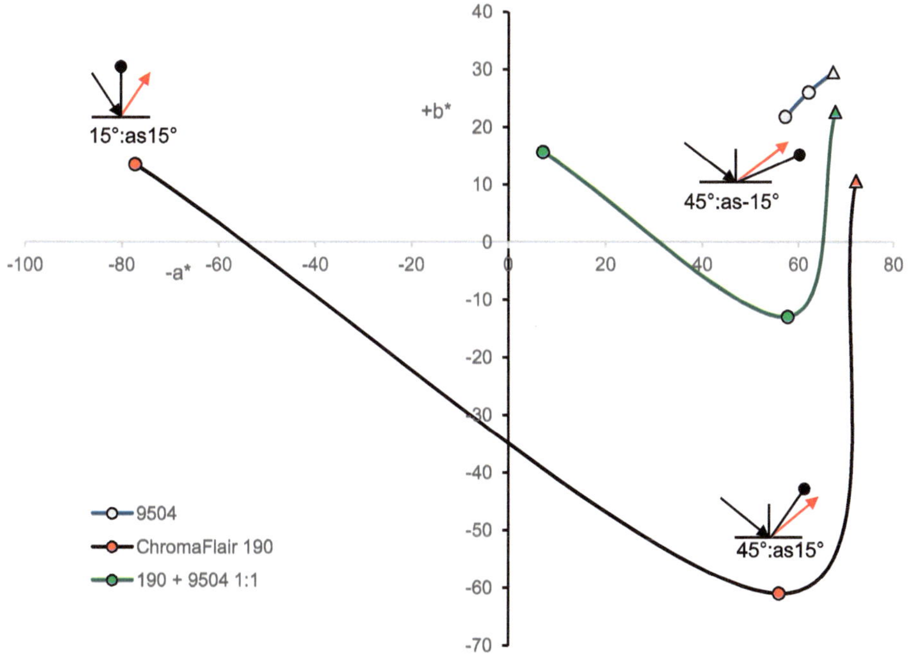

Figure 114: ChromaFlair blended with Red 9504.

shorter arm ranges from classic geometry at 45°:as15° to flat at 60°:as15° = 45°:as–15° (in measurements with the Zeiss GK311/M up to 25°:as15°). This form of interference line and the associated interference colors can be shifted almost as shown by the 1:1 mixtures with white 9103, yellow-gold 9303 (iron oxide/titanium dioxide), and red 9504 (iron oxide). The color character of ChromaFlair 190 is retained in the blends (Figure 112, Figure 113, Figure 114).

As with all blends, it is worth experimenting yourself. There are no limits to the choice of interference pigments.

Blends with aluminum pigments

Aluminum pigments differ in terms of their shape: cornflakes have irregular structures like flattened potatoes (hence the name), while silver dollars have a round structure and, as the name suggests, resemble dollar coins. When liquid aluminum is atomized in the normal way, irregular shapes are produced, which then become cornflakes. If this process is carried out under a protective gas such as argon, it results in spherical structures, which then become silver dollars.

Aluminum pigments – whether fine or coarse – reflect incident light extremely well. Close to specular at 15° and 25° aspecular, they reflect more than white interference pigments and are therefore significantly brighter than them. If a white interference pigment is mixed with aluminum in a paint, the paint becomes darker close to gloss.

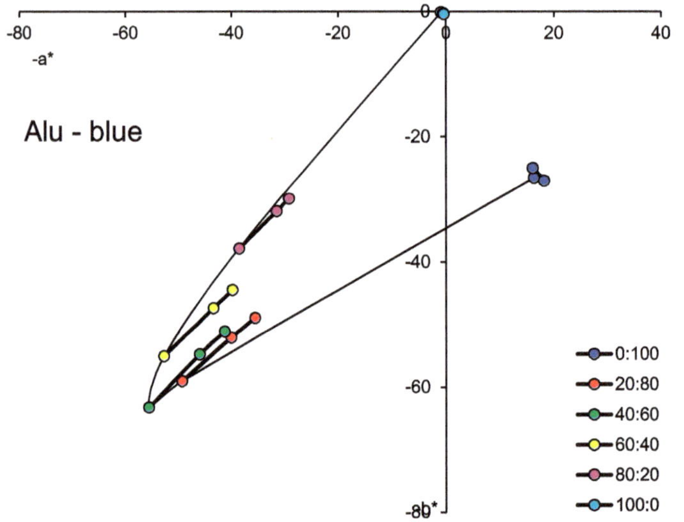

Figure 115: Blends of aluminum with blue pigment initially show an increase in chroma up to the turning point (green).

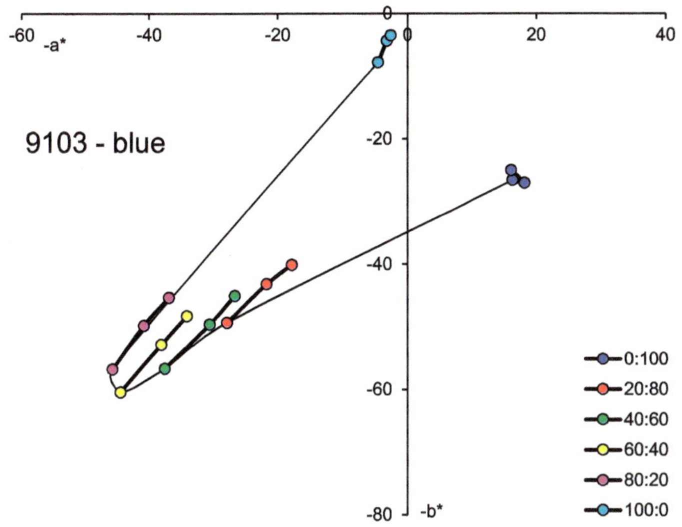

Figure 116: Blends of Perlwhite 9103 with blue pigment also initially show an increase in chroma up to the turning point at the turning point (yellow).

When aluminum pigments are illuminated more flatly, the reflection curves do not shift toward shorter wavelengths, in contrast to colorful interference pigments. For the corresponding a*b* values, the 45°:as15° to 45°:as–15° segment is an extension of the 45°:as25° to 45°:as15° segment. It does not "bend" clockwise like colorful interference pigments. This behavior of aluminum pigments is also evident in white interference pigments, where the reflection curves also do not shift toward shorter wavelengths.

When mixing aluminum pigments with interference pigments, the result shows whether the use of a colorful interference pigment is "worthwhile." In most cases, however, the use of white interference pigments is recommended.

Currently, both types of pigments – interference and aluminum pigments – are used together in most automotive paints.

White interference pigments generally behave like aluminum pigments, as shown in the examples (Figure 115, Figure 116).

Blends with colored pigments

Blends of interference pigments with colored pigments display the reflective color of the interference pigment at angles of approximately 20° to 30° from specular. If this were not visible, mixing with a colored pigment would not make sense.

However, the reflective color can be disrupted by an aluminum pigment, which is significantly brighter than the reflective color of the interference pigment when viewed close. This depends on the respective pigment proportions.

When mixed with colored pigments, interference pigments – both colored and white – behave like their mixtures with white on the one hand, but on the other hand, they produce interesting color impressions that cannot be created with aluminum and colored pigments.

When green and blue color pigments are mixed with white, the chroma and brightness of their blends initially increase until they reach a turning point. From this turning point, the color "migrates" toward white, its chroma decreases, and its brightness continues to increase (Figure 117). When white is added, yellow and red color pigments blend directly between the starting pigments.

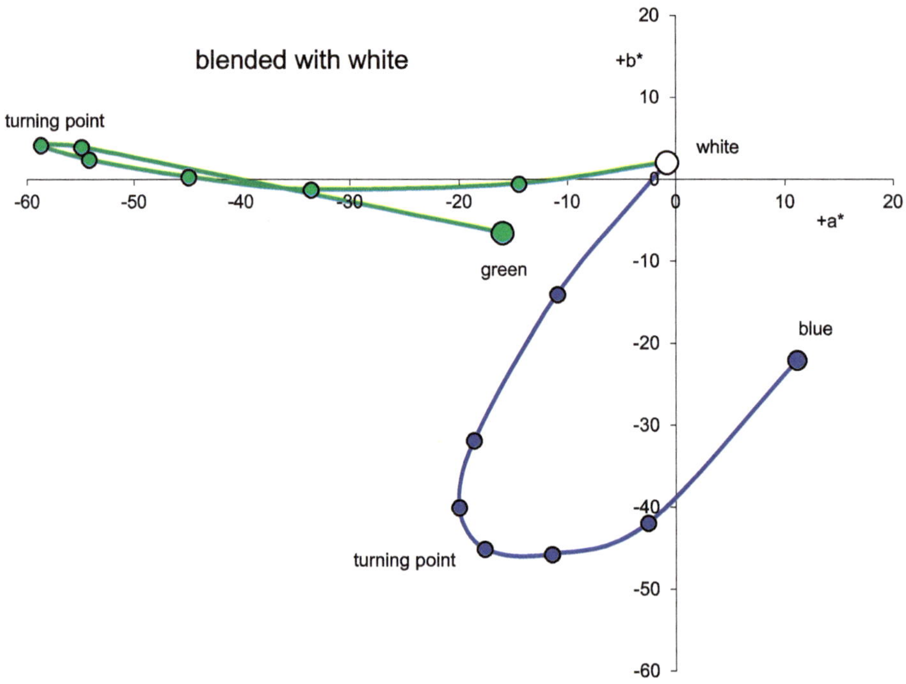

Figure 117: The typical behavior when mixing green and blue with white is also observed with interference pigments.

This behavior also occurs with blends of interference pigments – white and colored. Interference pigments initially make a blend with green or blue colored pigments appear more intense (= higher chroma) and brighter. At a turning point, the mixtures become less colorful and brighter.

Small additions of a colored pigment make the mixture more colorful up to a turning point (Figure 118, Figure 119). Here, small amounts of colored pigments act like tiny light emitters.

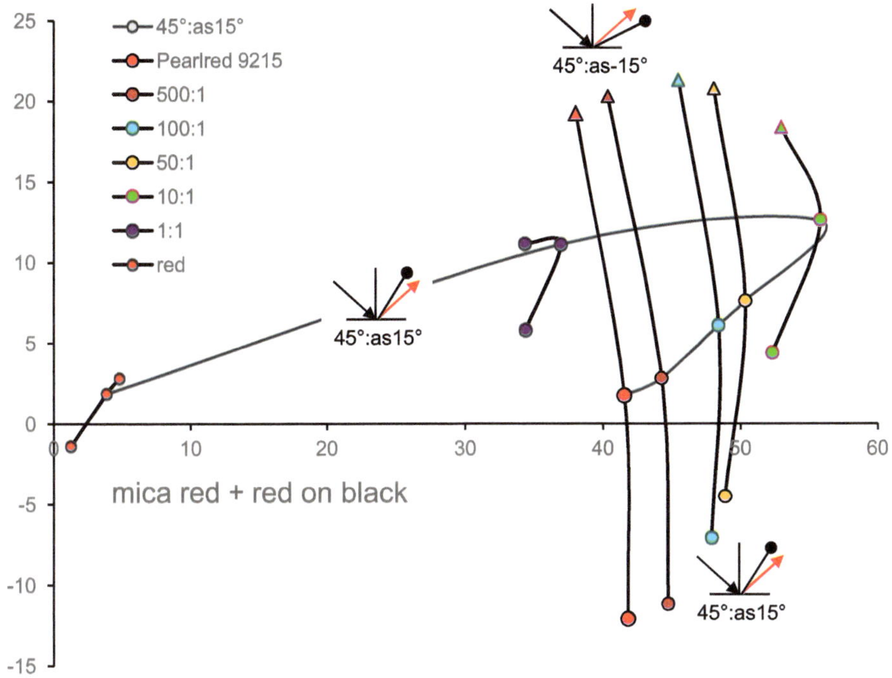

Figure 118: If you blend a little red into the red interference pigment 9215, the chroma initially increases.

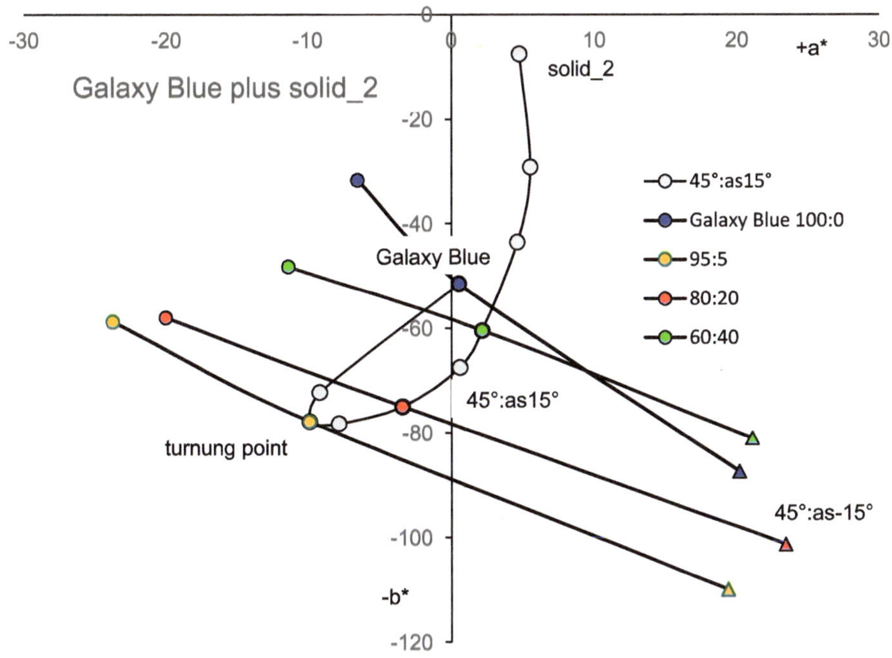

Figure 119: Galaxy Blue exhibits the same behavior with a blue pigment mixture.

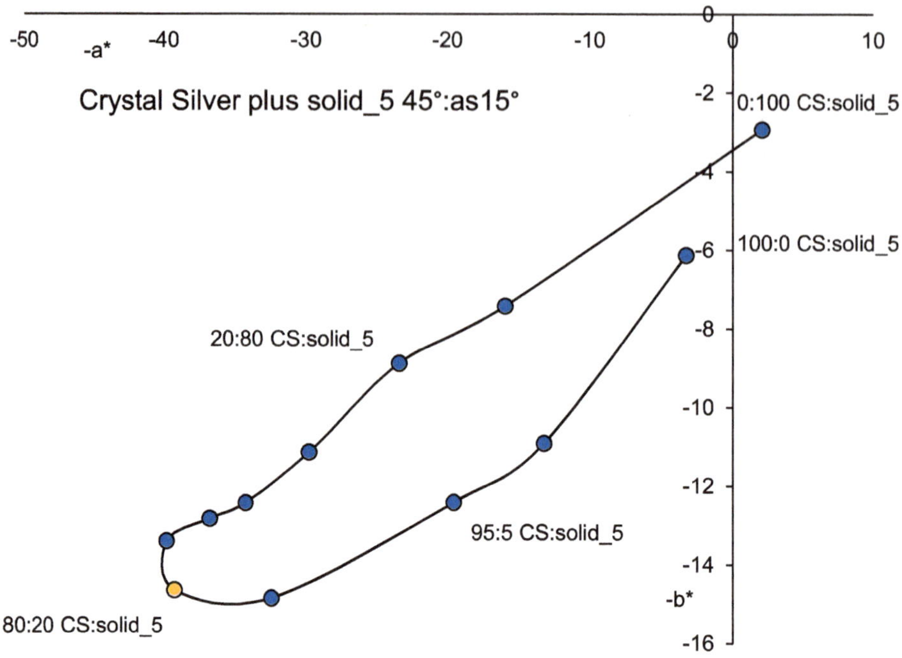

Figure 120: The mixtures exhibit the same properties as colored pigments.

Examples also include mixtures of Galaxy Blue and Crystal Silver with colored pigments (Figure 120). Up to a turning point, the mixtures become more colorful, which is also clearly shown in the chroma and brightness diagrams (Figure 121, Figure 122).

With larger quantities, the interference lines shift toward the colored pigment (Figure 123). The interference line remains visible up to a certain concentration and can be used for identification purposes. A green interference pigment and a red colored pigment do not mix like two color pigments. Close to the gloss, the interference pigment provides the color, while the color pigment provides the color after the transition area and further away from the gloss.

Colored pigments can be used to shift the position of the color in the color space. For example, white Crystal Silver and blue Galaxy Blue exhibit typical behavior when mixed with carbon black or a colored pigment (Figure 124).

Pigments with the same name are offered by different manufacturers. Since interference pigments are “individualists,” they cannot be exchanged at will (Figure 125).

The interference line plays an important role in identifying one or more interference pigments in a blend with colored pigments. Figure 126 shows two blends of the same series color – interference pigment + colored pigments – that were accepted by the car manufacturer. The interference line, which would confirm the visual differences, was not measured (Figure 127).

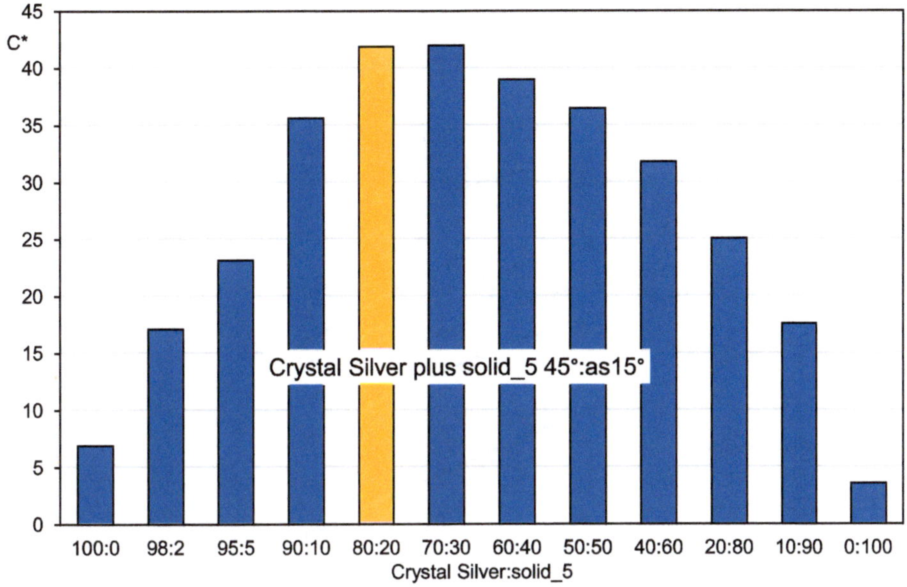

Figure 121: The color of the turning point (orange column) has the highest chroma.

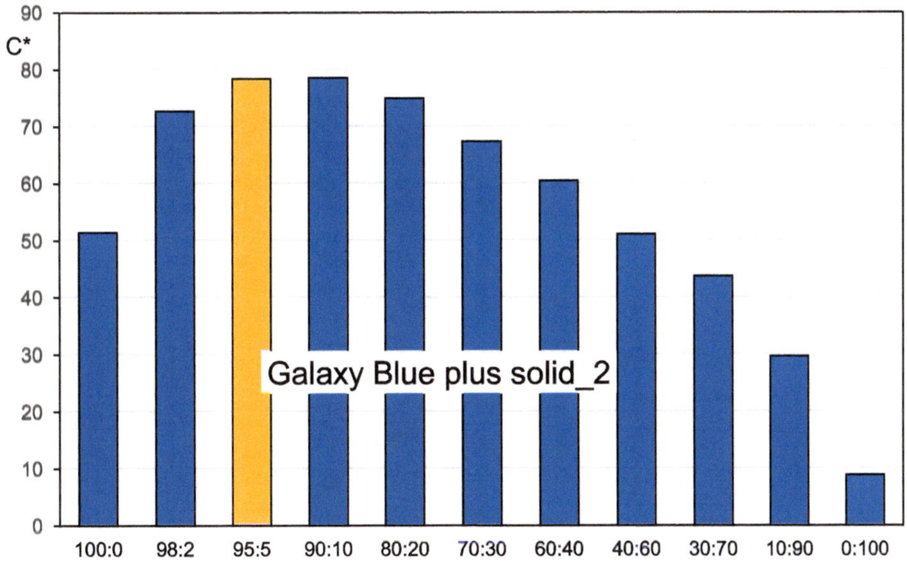

Figure 122: The turning point (orange column) has the highest chroma.

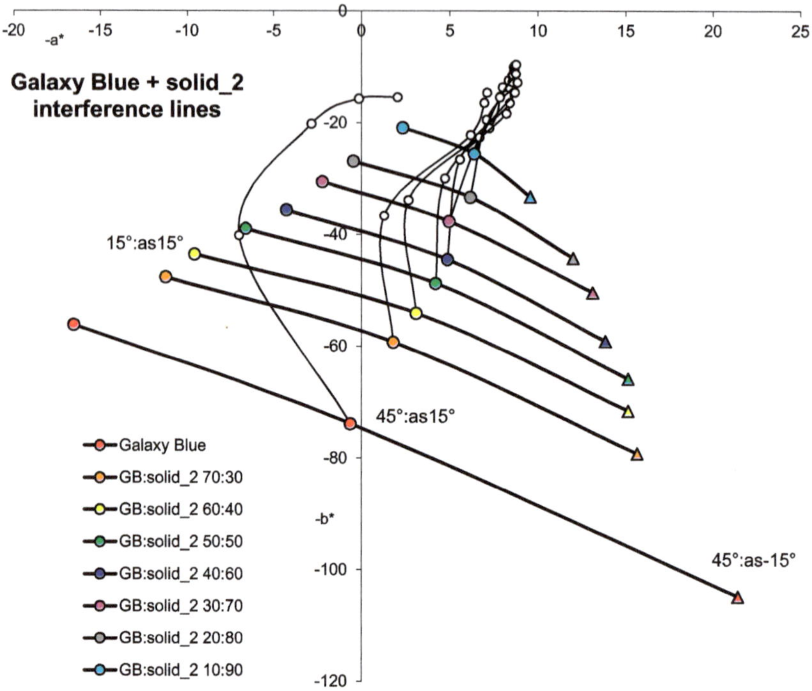

Figure 123: Combination of Galaxy Blue and color pigment.

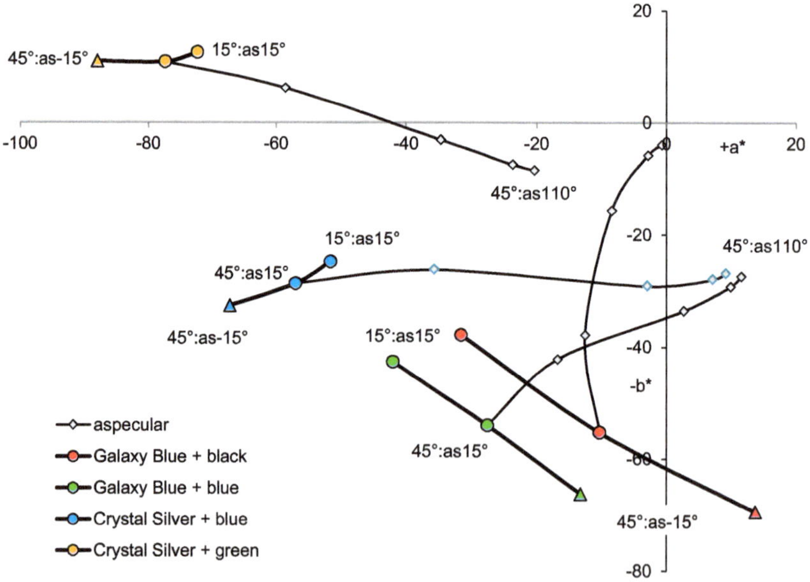

Figure 124: Blends with colored pigments.

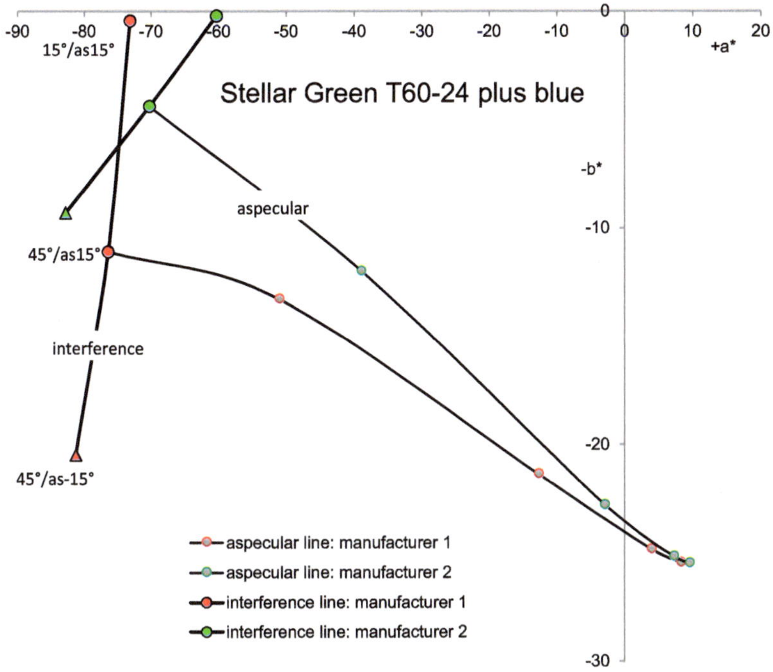

Figure 125: Interference pigments with the same name can have different color effects.

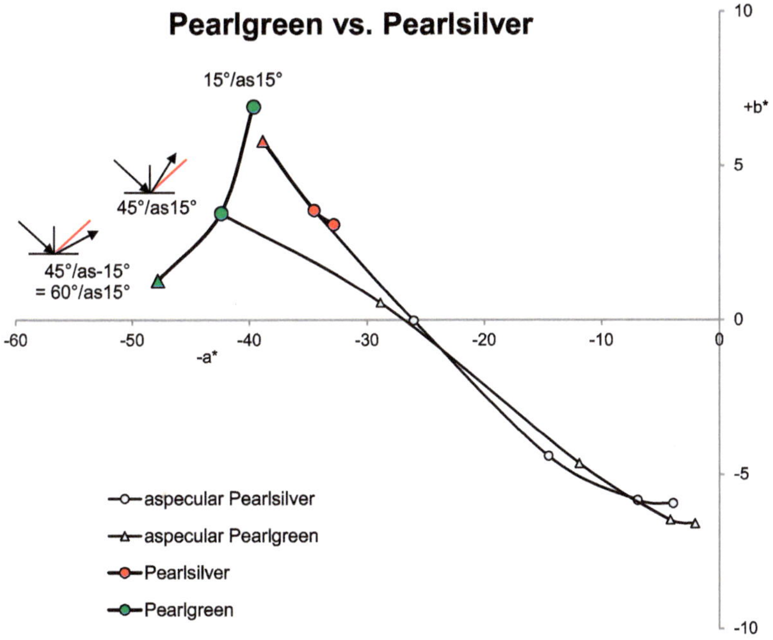

Figure 126: Typical example of the same car series color with different interference pigments: Here, different interference lines.

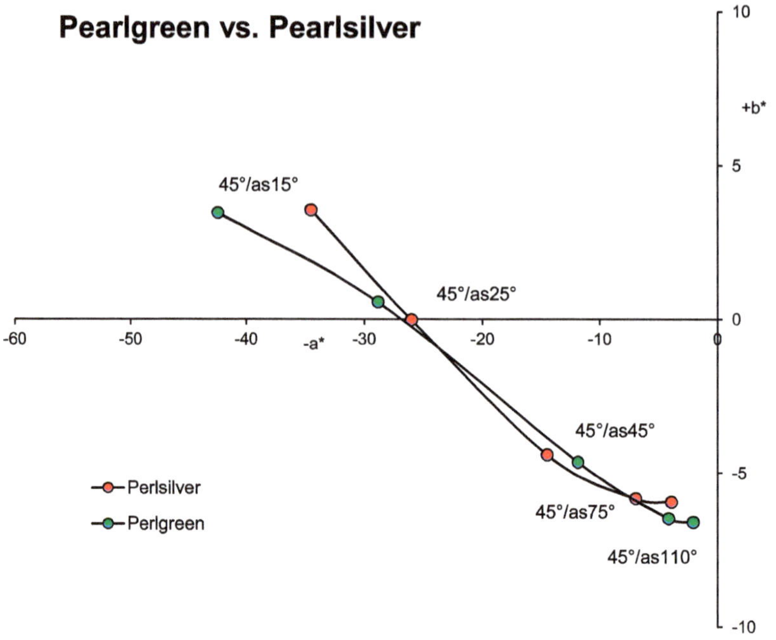

Figure 127: The representation without interference lines shows almost identical aspecular lines.

Blends with aluminum and colorful pigments

The "highlight" of the mixtures is that with the three types of pigments. These are often found in automotive paints, where the color is controlled by the colored pigments and the effect by the interference pigment, while the covering power is controlled by the aluminum pigment.

Car paints usually contain four to eight pigments. To simulate these, mixtures of colored paints were prepared. These were used as the basis for blends with interference and aluminum pigments. These blends were prepared in systematic steps.

The color reactions are shown in a*b* charts, with particular attention to the interference line. In the two starting positions – the mixture with aluminum pigment and the mixture with interference pigment – the familiar reactions of the interference lines are evident: the blend with aluminum shows the extension of the specular line, while the blend with the interference pigment reproduces the typical counterclockwise kink. The blends between these two initial mixtures – i. e., the blend of interference, aluminum, and colored pigments – reflect the changes in the interference lines. Due to their course, these interference lines indicate the presence of a colored interference pigment in the blends.

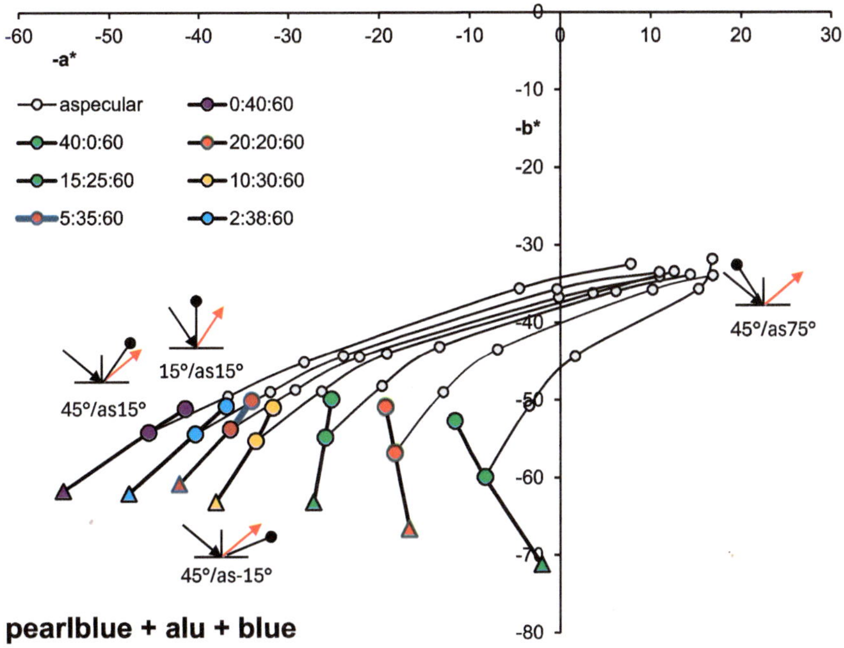

Figure 128: Change in the interference line when blending aluminum and Pearl Blue with blue pigment.

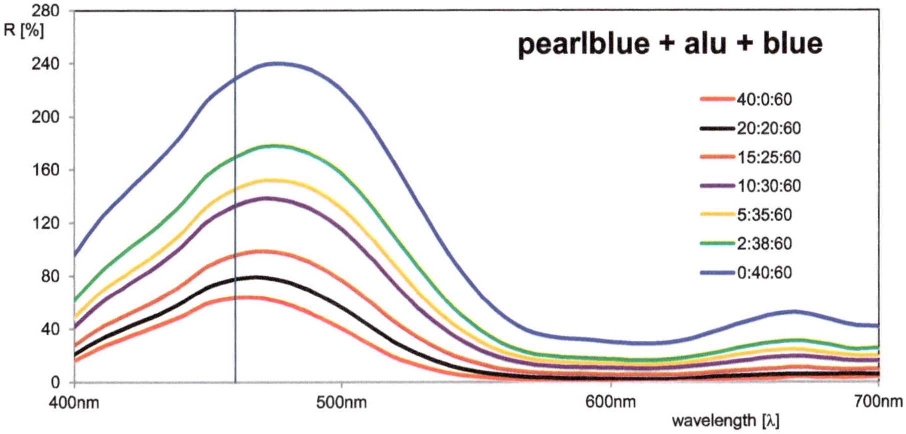

Figure 129: Reflection curves of blends of aluminum and Pearl Blue with blue color pigment.

An example is the mixture of Pearl Blue (9225) with blue color pigment and aluminum pigment with the same color pigment. The interference line of 9225 rotates clockwise in the three-color mixtures to the respective specular line (Figure 128). The reflection curves shift toward green due to Pearl Blue 9225 (Figure 129).

5 Visual and instrumental assessment

Wolf Moritz Cramer

Color assessment is one of the most important methods used in the color laboratory. Visual assessment, for example, at a window or in a light booth, plays just as important a role as instrumental measurement.

Visual assessment

In most cases, the object to be evaluated is observed and assessed without the exact geometries of the illumination and observation being known. This applies to all areas of printing processes, plastics manufacturing, and automotive paints.

Our eyes are the most important tool for assessing colors. This is where the light rays are reflected by objects. In the retina of the eye, they encounter cones that are sensitive to different spectral ranges (Figure 130). The cones convert the light rays into optical stimuli, which are transmitted to the brain. There they are "translated" into colors. There are no colors in front of our eyes, only light rays (= physics). Colors are only created in the brain.

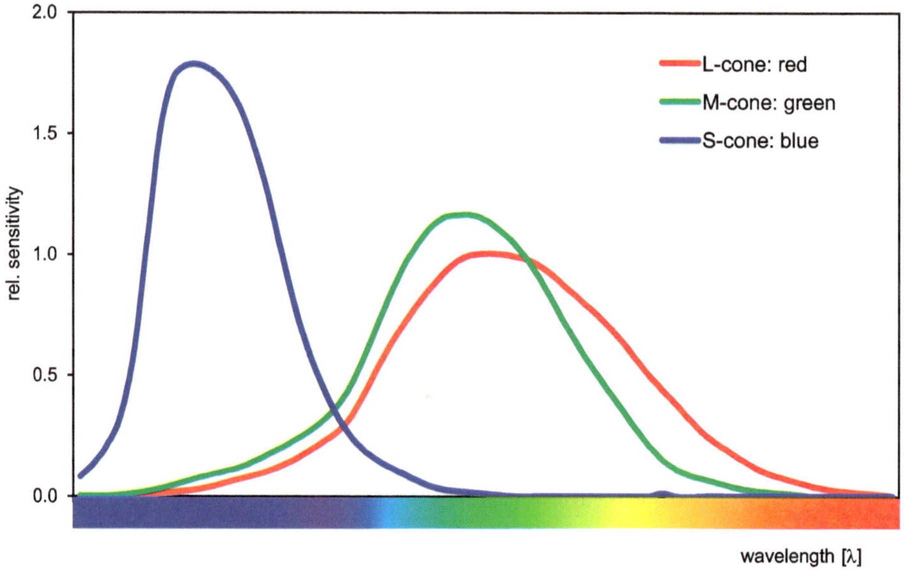

Figure 130: Sensitivity curves of the three cones LMS.

Since the cones for the green and red spectral ranges are so close together, we cannot see dark yellow. We know dark red, dark green, and dark blue, but not dark yellow.

https://doi.org/10.1515/9783111439204-005

In complementary color theory, yellow and blue are opposites, as are red and green (Figure 131), which is also in line with our perception.

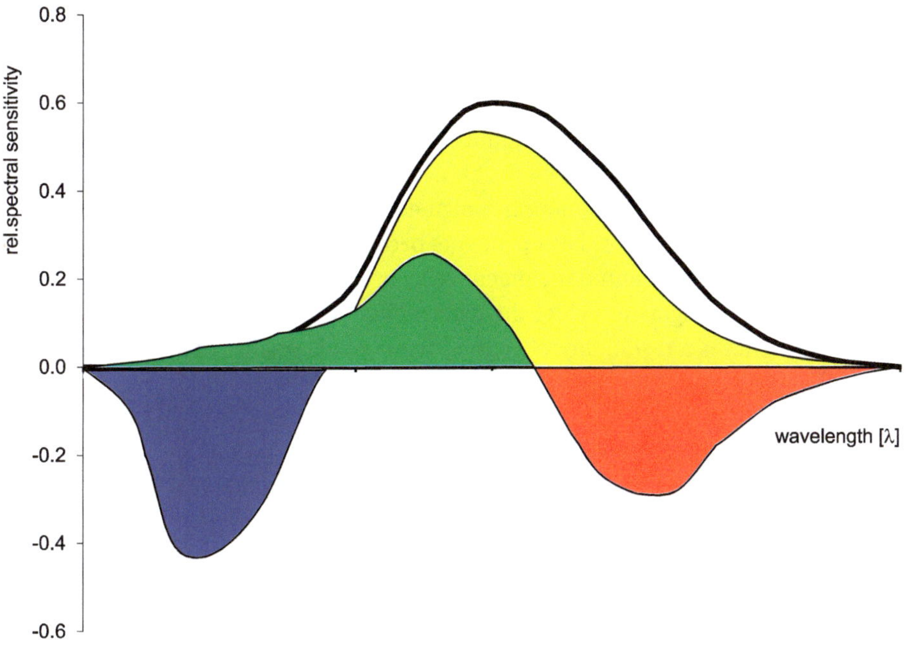

Figure 131: Representation of opponent colors.

When evaluating sample panels coated with effect pigments, they are usually moved up and down at a window or in a light booth. The aim is to observe and evaluate them under different geometries. Geometry refers to the combination of illumination and viewing angles. The normal is perpendicular to the sample panel.

When performing a visual assessment at a window or in a light booth, the geometries are usually unknown. However, they are just as important as in an instrumental measurement to precisely define and describe the color gradient and effect. For this reason, we have developed the GonioViewer, which precisely reproduces geometries. In addition, the observation area is approximately the same size as the measuring area of the devices. It should also be noted that when assessing a panel sized like a postcard with your arm outstretched, there is a viewing angle of up to 20° between the upper and lower edges. This is greater than the near-gloss geometries at 15° and 25° (Figure 132). This can result in differences between visual and instrumental assessments. With GonioViewers, these differences are avoided due to the size of the observation.

The different observation and measurement variables lead to the curious phenomenon that we visually perceive a panel with a white interference pigment as

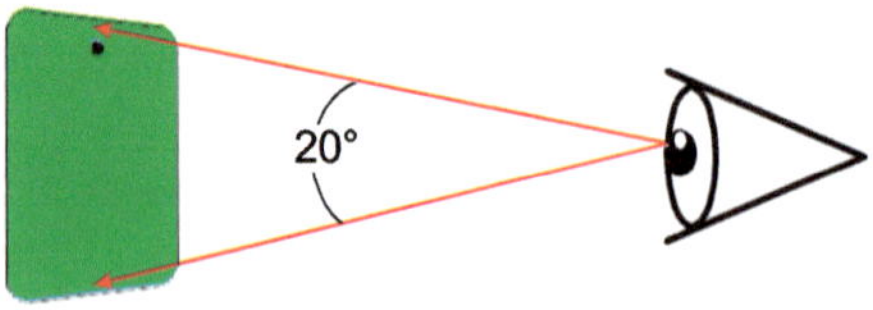

Figure 132: The viewing angle between the top and bottom edges of a panel can be up to 20°.

brighter than a panel with an aluminum pigment (Figure 133). However, the measurement shows that the panel with the aluminum pigment is brighter, especially near specular (Figure 134). If the aluminum pigment is blended with the white interference pigment, the mixture of both pigments becomes darker rather than lighter. The reason for this is that aluminum pigment reflects almost all incident light rays, while white interference pigment divides the incident light into a reflection and a transmission component. As a result, the reflection component is smaller than with aluminum pigment.

Figure 133: Right: white, Left: aluminum.

During visual assessment, two or more panels can be examined at the same time, whereas during instrumental assessment, the panels are measured one after the other.

Various sampling scenarios are presented, which we have recreated using GonioViewers. The GonioViewers are 3D-printed and measure 24 cm long, 12 cm high, and 3.5 cm wide. The device is semicircular in shape and is placed lengthwise on one or two sample sheets. The semicircular arc has openings for tubes that point inward toward the panel. It is illuminated by a lamp that is inserted into an opening via an adapter. By looking successively through the openings, you can follow the color and brightness gradient.

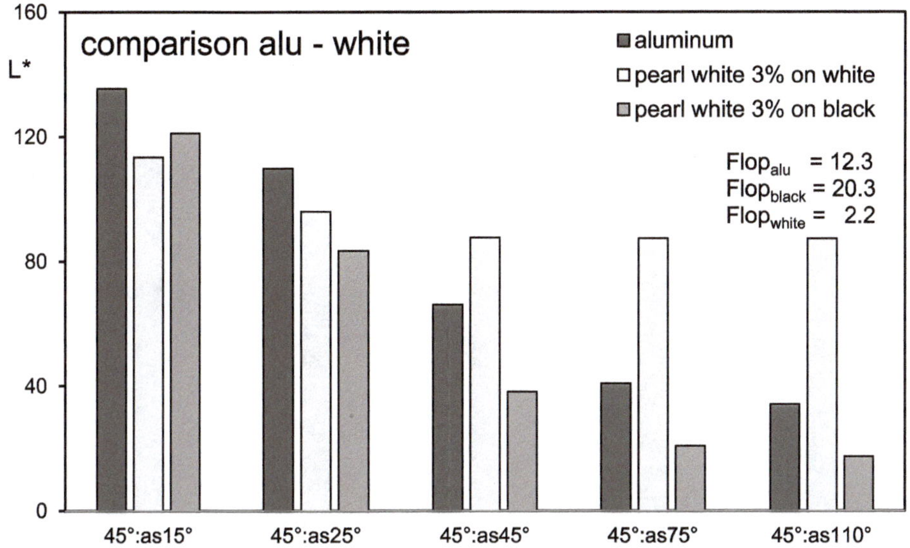

Figure 134: Brightness changes with white interference pigment on white and black substrates compared to aluminum.

GonioViewer 01

Newer interference pigments display a rainbow when illuminated at 45°, for example. This begins at an aspecular angle of approximately 25° from the corresponding specular. Up to this aspecular angle, the pigments appear gray or colorless. From 25°, they initially start with blue and change color to green at 45° from the gloss. At an aspecular angle of approximately 55° from the gloss, they change to yellow and, further away from specular to red. Figure 135 and Figure 136 show different perspectives: close to specular, no colors can be seen, but at 45° aspecular the variety of colors becomes clear. In Figure 137, this color behavior is illustrated on both sides of the normal line for curved panels.

The comparative color measurement shows the limited representation of the color gradient with the current geometries (red line) compared to measurements with the Zeiss GK311/M in 5° increments from specular (green dots) (Figure 138). The color shift can also be clearly seen in the reflection curves from 45°:as25° to 45°:as70° (Figure 139): Reflection at 45° illumination and 25° aspecular starts out achromatic. With a larger viewing angle from the gloss (aspecular), the maximum shifts from the UV range into the visible spectral range. The maximum is reached at 45°:as40°. From 45°:as70°, it becomes achromatic again.

To visually observe this color gradient, this GonioViewer is equipped with observations at –15°, 15°, 25°, 35°, 45°, 55°, 65°, 75°, and 110°. The angles 35°, 55°, 65° are not available on current multiangle devices. They cannot display the color gradient from blue to green and yellow, which goes directly from blue-green (as45°) to red (as75°) (Figure 140).

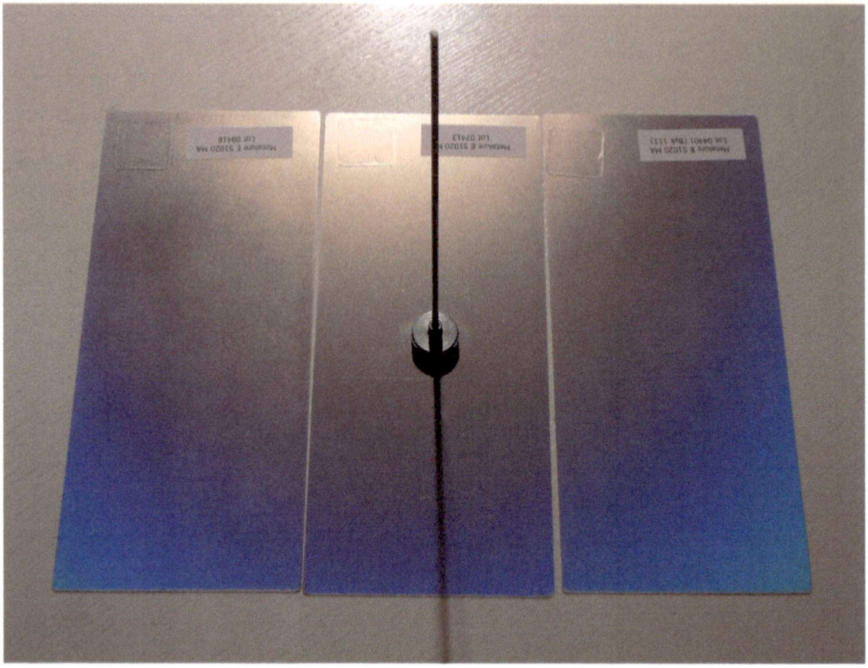

Figure 135: Prismatic aspecular.

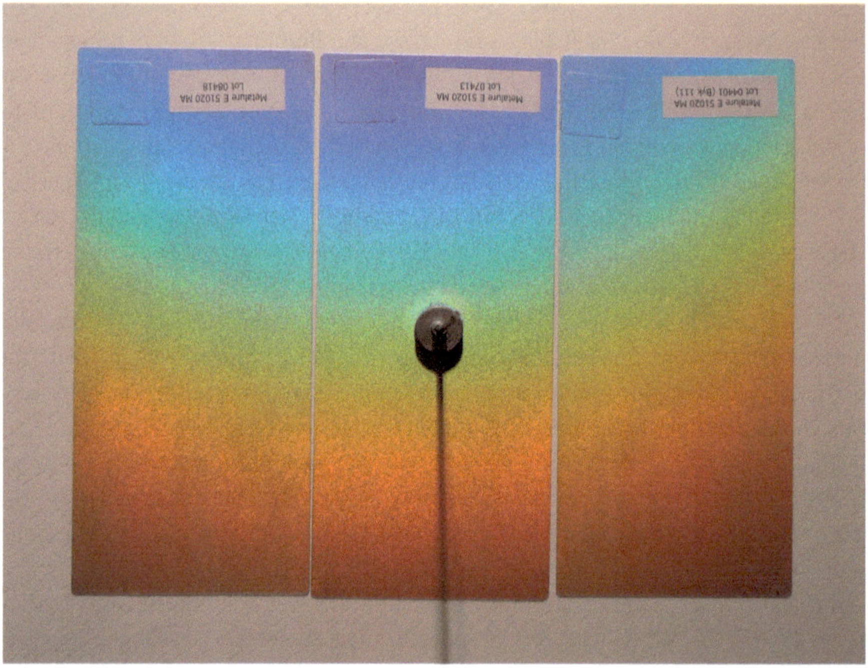

Figure 136: Prismatic closer to specular.

Figure 137: Diffractive pigments show colors from 25° aspecular.

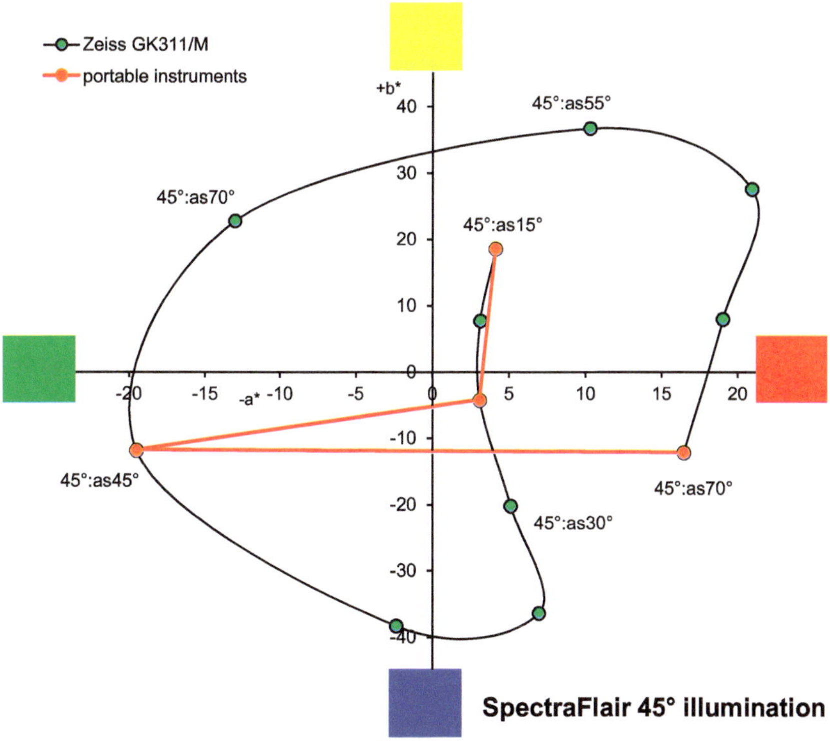

Figure 138: Measurement with the Zeiss GK311/M compared to a portable device.

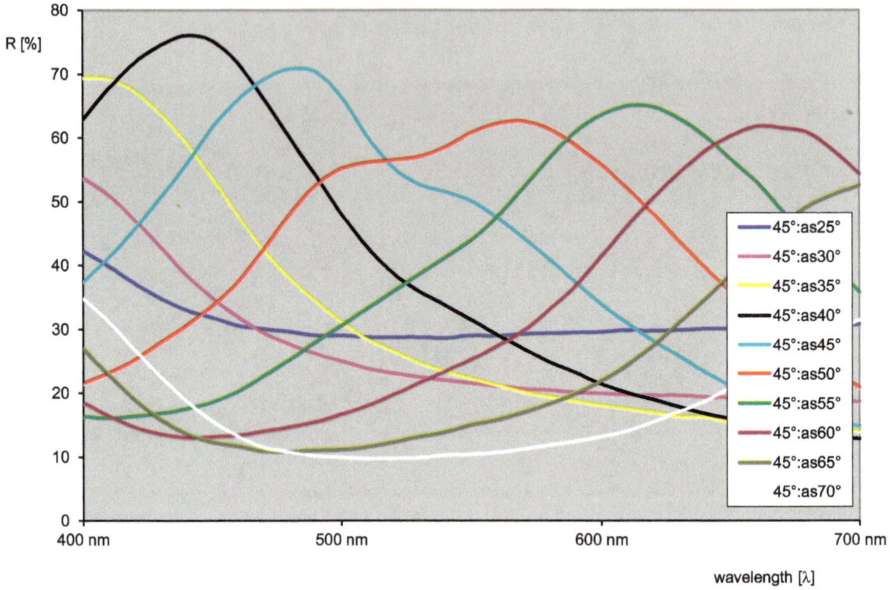

Figure 139: The reflective color shifts from achromatic to blue, green, and then to yellow and red.

Figure 140: GonioViewer 01.

GonioViewer 02

When observing one or two panels at the window or in the light booth, these are held in the starting position so that the light source is reflected in them or the highest gloss is achieved.

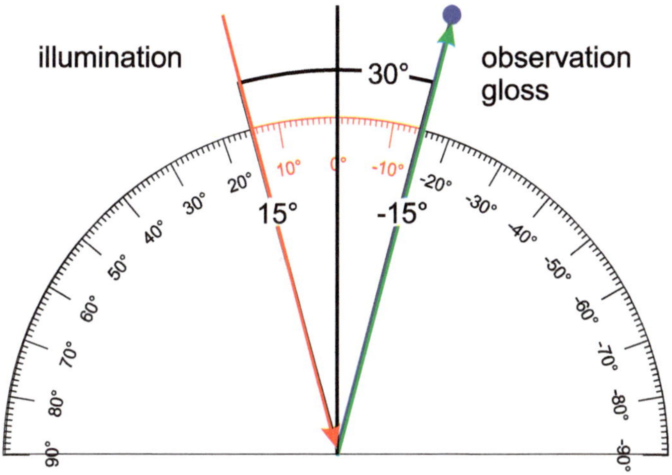

Figure 141: Initial situation during observation at the window: illumination at 15°, observation 30° angle of difference to illumination.

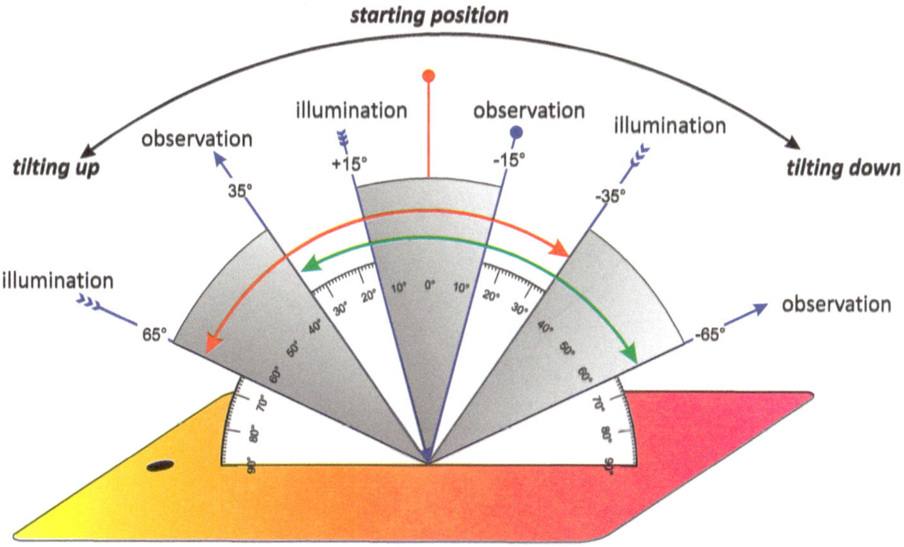

Figure 142: Geometries when moving the panel up and down.

If an illumination angle of 15° is used in this initial position, the observation angle is –15°. The normal is perpendicular to the panel (Figure 141). The angle between illumination and observation is therefore 30° (Figure 142). This angle remains the same throughout the entire observation process.

Tilting the sample sheet downwards

If the panel is tilted downward – away from the viewer – the normal also moves away from the viewer, but the illumination angle moves toward the viewer:

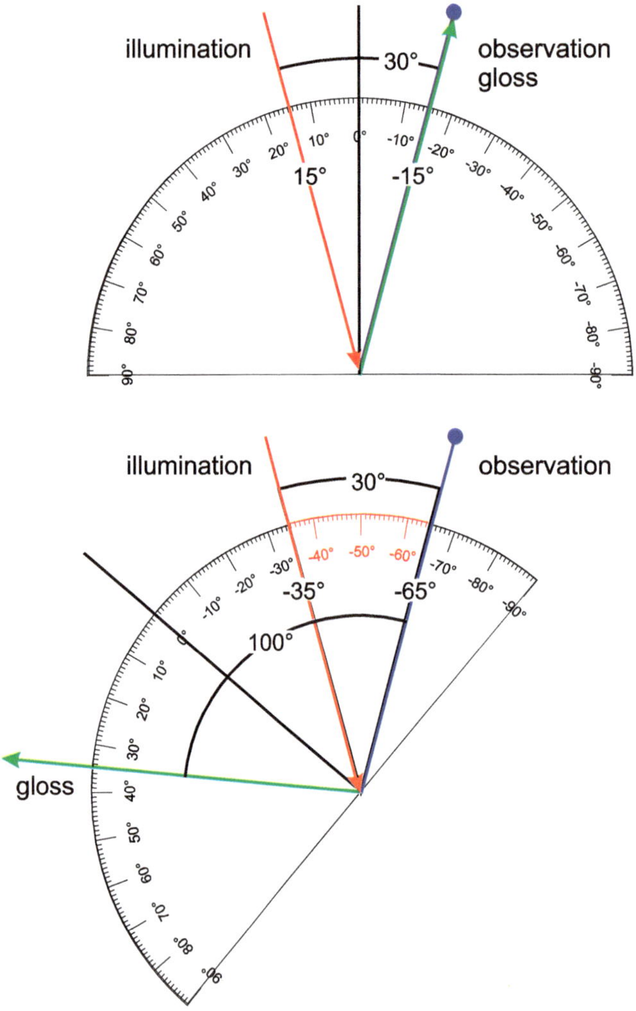

Figure 143: Geometries when moving downwards.

The light source (illumination) moves from 15° to 5° and the observer to –25° (the 30° difference between illumination and observation always remains the same). With illumination of 5°, the gloss angle is –5°, i. e., the observer at –25° is then –20° aspecular. If the illumination moves to the normal, the observation is at –30°, which corresponds to an aspecular angle of –30°.

If the panel is tilted further, the illumination switches to the side of the normal on which the observation is taking place (Figure 143). For example, at –35° observation, the gloss angle is +35°. The observation then takes place at –65° with a difference angle of +100°. As the tilt increases, the illumination angle increases from –35° to –55° and more. Accordingly, the difference angle (aspecular) increases from +100° to 140°.

Tilting the sample sheet upwards

If the observer moves the panel toward him, the angle of illumination and, accordingly, the angle of gloss also change. For all movements, the angle of observation between the angle of illumination and the respective angle of gloss always remains the same. For example, if the angle of illumination changes to 65° from normal, the corresponding angle of gloss is –65°. The angle of observation is then +35° (= 30° difference between illumination and observation) and the difference angle between observation and gloss is accordingly +100° (Figure 144).

The optical law states that the illumination angle and the angle of reflection can be swapped (law of light reversal). If we compare the geometries of the processes described, taking light reversal into account, the observer assumes the same geometries when tilting up and down.

For this reason, only a series of geometries were implemented in GonioViewer 02. The starting position is illumination at 15° with observation at –15°. The illumination and observation are then changed in 10° increments: illumination from 15° to 25° and observation from –15° to –5°, illumination further to 35° and observation to 5°. The difference between illumination and observation remains constant at 30° (Figure 145).

With the GonioViewer 02, a starting position with illumination at 25° and observation at –25° can also be adopted. The angular positions are then changed accordingly.

Like all GonioViewers, the GonioViewer 02 is designed for comparative observation of two panels placed side by side.

It should also be noted that no geometry corresponds to the geometry of current multiangle devices when tilting a panel up and down at the window or in a light booth. This is a particular advantage of the GonioViewer 02.

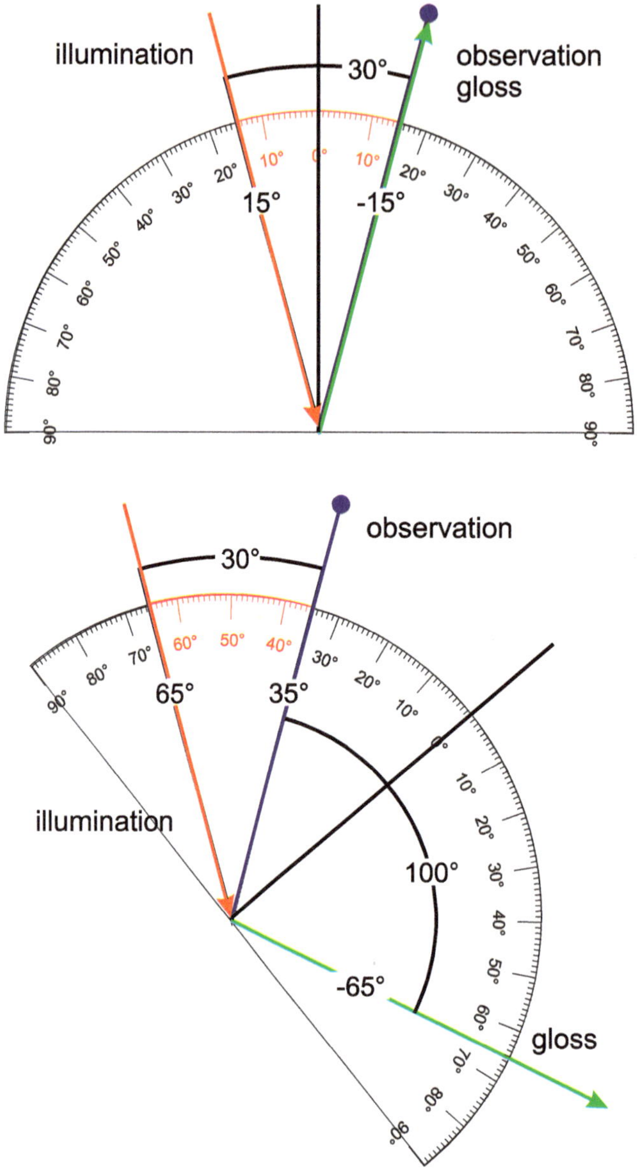

Figure 144: Geometries when moving upwards.

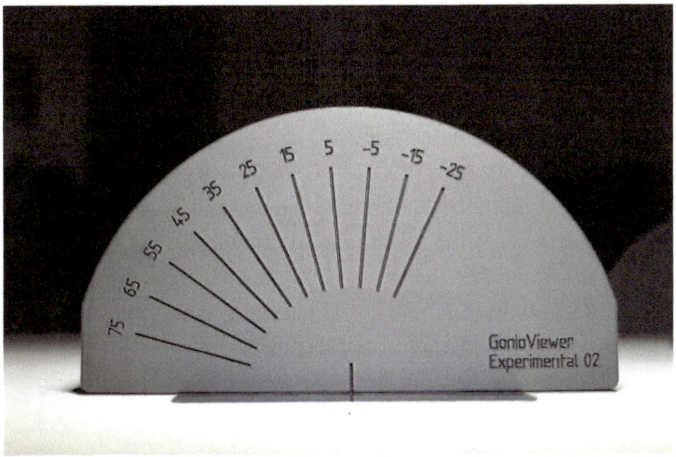

Figure 145: GonioViewer 02.

GonioViewer 03

The portable multiangle devices illuminate at 45° and measure at –15°, 15°, 25°, 45°, 75°, and 110° from the corresponding gloss angle at –45°. Based on my years of testing, the –15° difference angle has been adopted as the standard method E2539 of the ASTM. The other geometries were first introduced in 1995 as a portable device. They resulted from the combination of the difference angles of 15°, 45°, and 110°, which were proposed by DuPont, and the difference angles of 25°, 45°, and 75°, which were based on an idea by Zeiss.

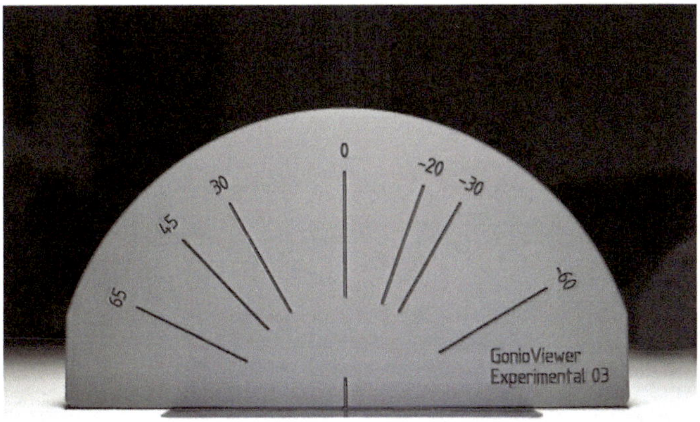

Figure 146: GonioViewer 03.

The results of a measurement are difficult to visualize. As mentioned above, the geometry of a panel tilted up and down at the window does not correspond to the geometry of an instrument. With the GonioViewer 03, panels can be observed under the same geometries as with an instrument (Figure 146).

Two advantages of the GonioViewer: with this GonioViewer, two panels can be compared directly side by side under all geometries. And with the GonioViewer, observations are made with the same precision as with the instrument. When observing a panel normally, the difference between the upper and lower edges of a sheet (postcard size) can be up to 20°. This difference is greater than the difference between the first two near-gloss geometries of the portable measuring instruments at 15° and 25°.

GonioViewer 04

The resulting color of an interference pigment and a corresponding paint or plastic depends on the angle of the incident light: the flatter the illumination, the more the resulting color shifts toward shorter wavelengths (Figure 147). The angle of difference between specular and observation remains constant: red becomes yellowish, yellow becomes greenish, green becomes bluish, and blue becomes reddish (Figure 148).

No portable device allows for multistage adjustment of the illumination angle. Even when observing at a window or in a light booth, the illumination angle and the difference angle are changed.

A "correct" visual observation is made with the arm outstretched and the panel resting on the hand. In this position, the panel is illuminated flat and observed flat. The

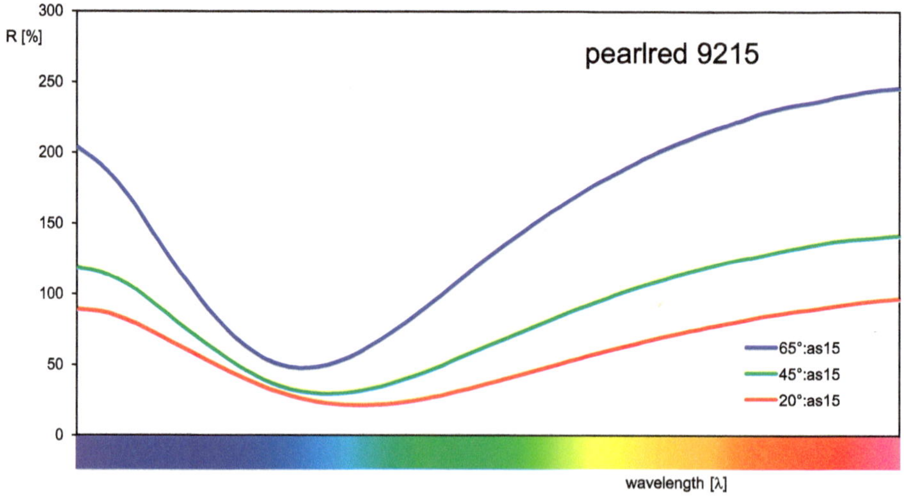

Figure 147: Shift toward shorter wavelengths with flatter illumination.

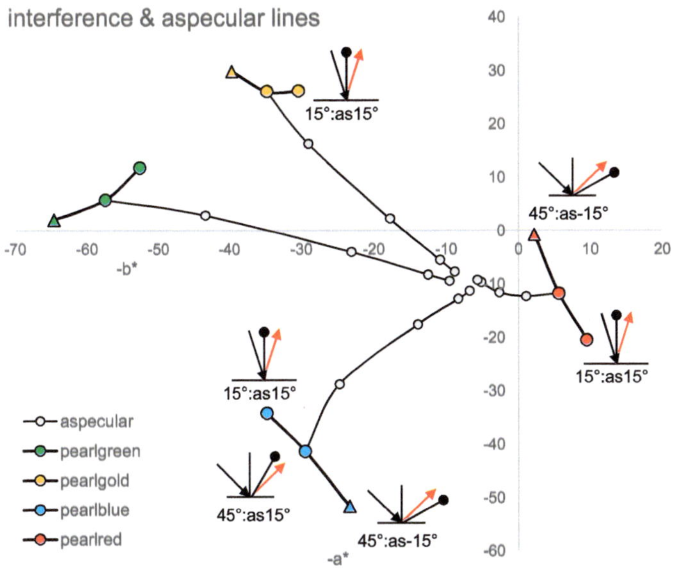

Figure 148: Interference lines of different interference pigments.

arm is then lowered while the panel is moved parallel downwards on the hand, where it is illuminated steeply and observed (Figure 149).

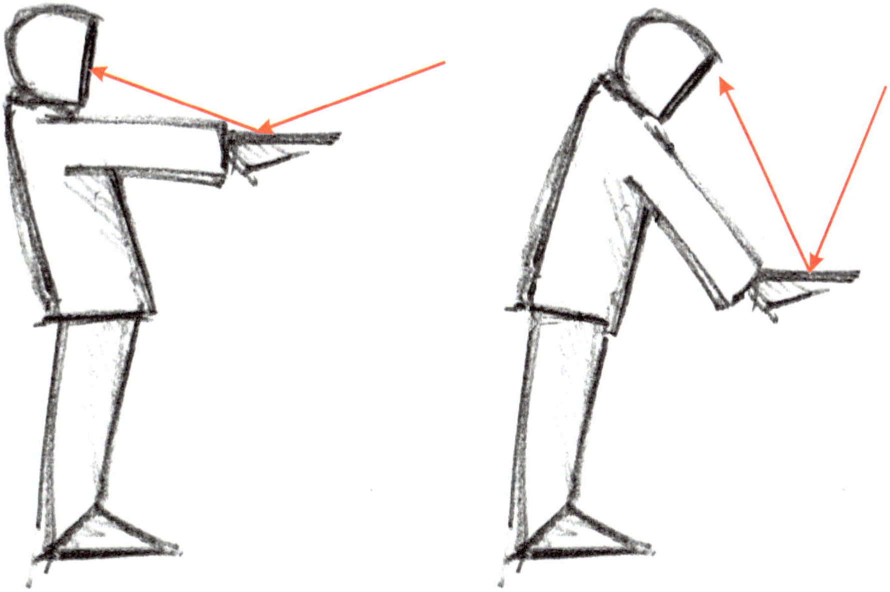

Figure 149: Flat and steep observation.

The GonioViewer 04 is suitable for observing color shifts due to interference. The starting position is steep illumination at 15° from the normal and observation at 0° (=15° difference angle from the gloss angle at −15°). Then change the illumination to 25° and observe at −10° (= 15° angle of difference). Continue this series up to a flat illumination at 65° and observation at −50° (Figure 150).

Figure 150: GonioViewer 04.

In this series, color changes in interference can be observed in connection with changes in the angle of illumination.

With this GonioViewer 04, too, two sample sheets can be observed side by side.

GonioViewer 05

This GonioViewer is suitable for visual inspection of the geometries of the BYKmac with its Sparkle geometries and the X-Rite MA-T12 with its second illumination at 15° (Figure 151).

The BYKmac also records the sparkle parameter: Here, a black and white camera takes normal images of a panel that is illuminated at 15°, −45°, and −75°. These geometries are also integrated into GonioViewer 05.

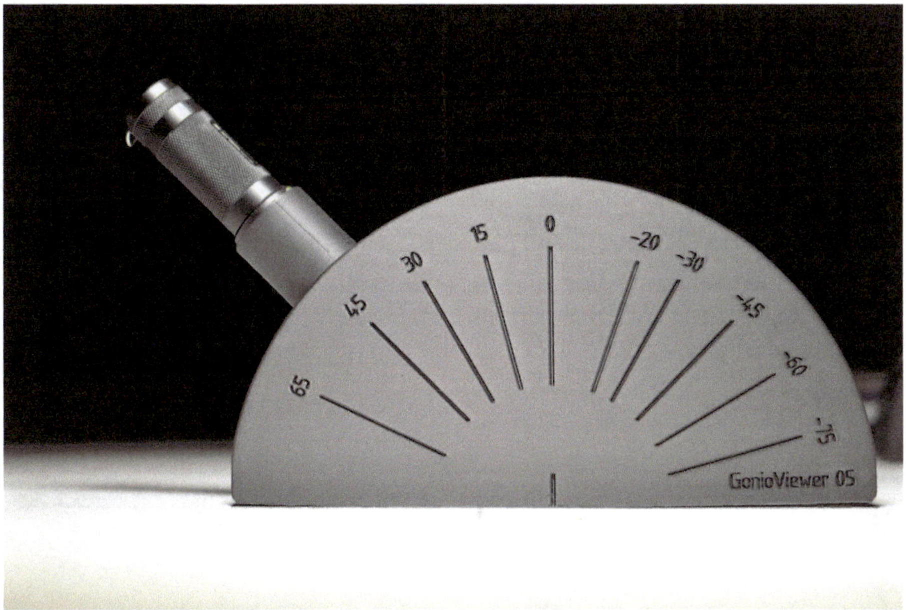

Figure 151: GonioViewer 05 with lamp.

The MA-T12 from X-Rite measures at 45° and illuminates at –15°, 15°, 25°, 45°, 75°, and 110° from the gloss angle at –45°. A second measuring angle at 15° measures at the same positions, resulting in difference angles of 15°, 45°, and 80° as well as –5°, –15°, –30°, and –45°. X-Rite offers these geometries for sparkle and graininess determination.

All the geometries mentioned can be reproduced for visual inspection using the GonioViewer 05. The GonioViewer 05 also offers the option of inspecting two panels side by side.

GonioViewer 06

Three lights are integrated into the BYKmac to determine the sparkle. A camera captures black and white photos for calculation.

Due to structural requirements, the lights are positioned at 15° from the normal as well as at –45° and –75°.

These three geometries are implemented in this GonioViewer 06 for visual inspection. Additional geometries are also integrated so that when observing in the normal (corresponding to the camera view) at 15°, 30°, 45°, 60°, and 75°, both sides of the normal can be illuminated. For the test, the position of the lamp is adjusted incrementally using the adapter (Figure 152).

Figure 152: GonioViewer 06 with lamp.

Aspects of visual assessment

The colorimetric comparison of the different samples shows the different color gradients of the visual processes. A panel is moved up and down at a window or in a light booth. Due to the law of light reversal, the geometries correspond to each other (Figure 153). This is also evident in the chroma C*, where the illumination angle or the aspecular angle changes at the same difference angle of 30° between the light source and the observer (Figure 154, Figure 155).

Plotting the interference line and the aspecular line together with the visual lines gives an overall picture of the pattern. In Figure 156, the corresponding goniometers for Stellar Green are also listed. This illustration also shows Figure 157 for Pearl Red. Figure 158 presents the visual samples as sketches of the geometries.

One aspect of visual sampling should be mentioned: if a panel or sample is placed on a table and is assessed by people of different heights, these people may have different color impressions. The taller person observes the panel from a steeper position than the shorter person. If this were to be transferred to the geometry of a measuring device, the taller person would see at as45° or as 25°, while the shorter person would see at as15°.

tilting down				
Illumination	specular	observation	difference	aspecular
15°	-15°	-15°	30°	0°
25°	-25°	-5°	30°	20°
35°	-35°	5°	30°	40°
45°	-45°	15°	30°	60°
55°	-55°	25°	30°	80°
65°	-65°	35°	30°	100°
75°	-75°	45°	30°	120°
25°	-25°	-25°	50°	0°
35°	-35°	-15°	50°	20°
45°	-45°	-5°	50°	40°
55°	-55°	5°	50°	60°
65°	-65°	15°	50°	80°
75°	-75°	25°	50°	100°

tilting up					reverse: illumination <> observation (exchange +-)				
Illumination	specular	observation	difference	aspecular	Illumination	specular	observation	difference	aspecular
15°	-15°	-15°	30°	0°	15°	-15°	-15°	30°	0°
5°	-5°	-25°	30°	-20°	25°	-25°	-5°	30°	20°
-5°	5°	-35°	30°	-40°	35°	-35°	5°	30°	40°
-15°	15°	-45°	30°	-60°	45°	-45°	15°	30°	60°
-25°	25°	-55°	30°	-80°	55°	-55°	25°	30°	80°
-35°	35°	-65°	30°	-100°	65°	-65°	35°	30°	100°
-45°	45°	-75°	30°	-120°	75°	-75°	45°	30°	120°
25°	-25°	-25°	50°	0°	25°	-25°	-25°	50°	0°
15°	-15°	-35°	50°	-20°	35°	-35°	-15°	50°	20°
5°	-5°	-45°	50°	-40°	45°	-45°	-5°	50°	40°
-5°	5°	-55°	50°	-60°	55°	-55°	5°	50°	60°
-15°	15°	-65°	50°	-80°	65°	-65°	15°	50°	80°
-25°	25°	-75°	50°	-100°	75°	-75°	25°	50°	100°

Figure 153: List of geometries and reverse geometries.

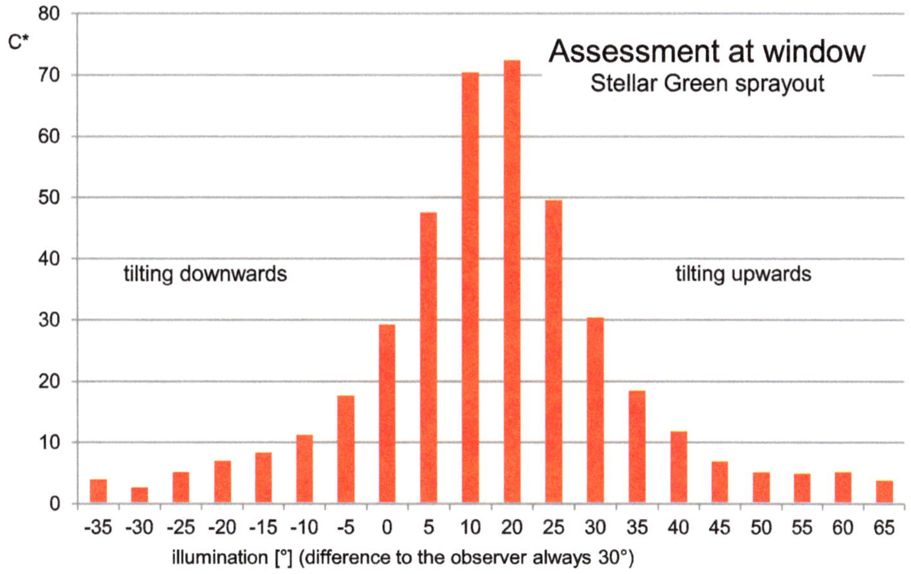

Figure 154: Stellar Green: Change in chroma when moving up and down at the window.

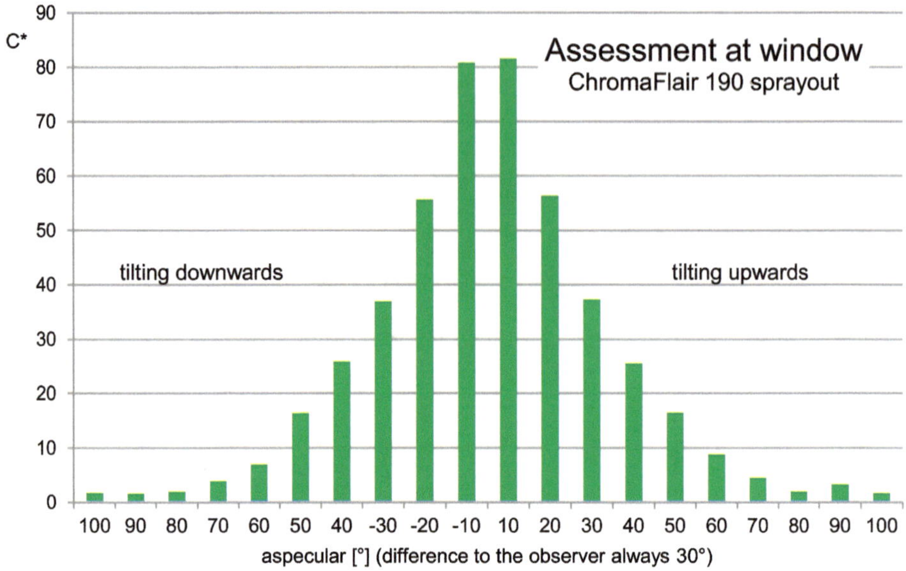

Figure 155: Chromaflair 190: Change in chroma when moving up and down at the window.

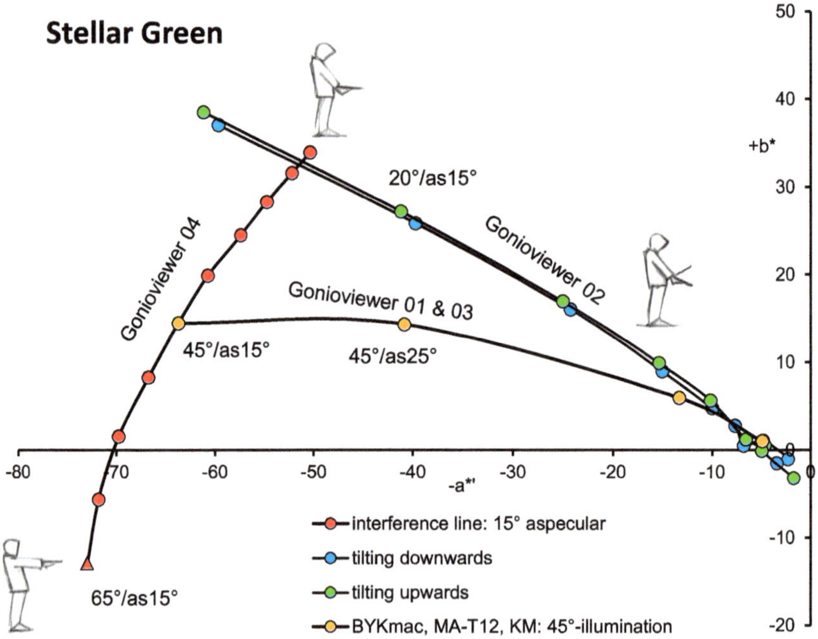

Figure 156: Comparison of the various observations and measurements for Stellar Green.

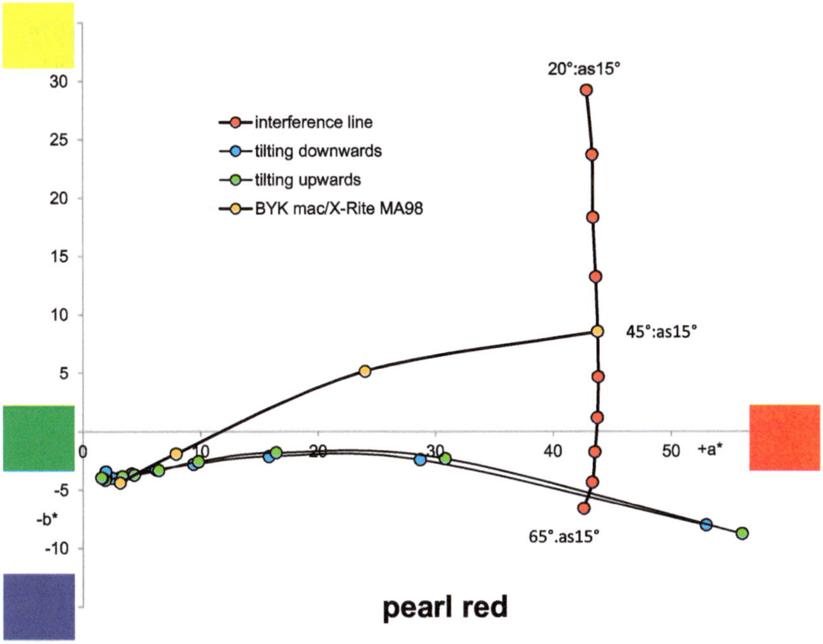

Figure 157: Comparison of the various observations and measurements for Pearl red.

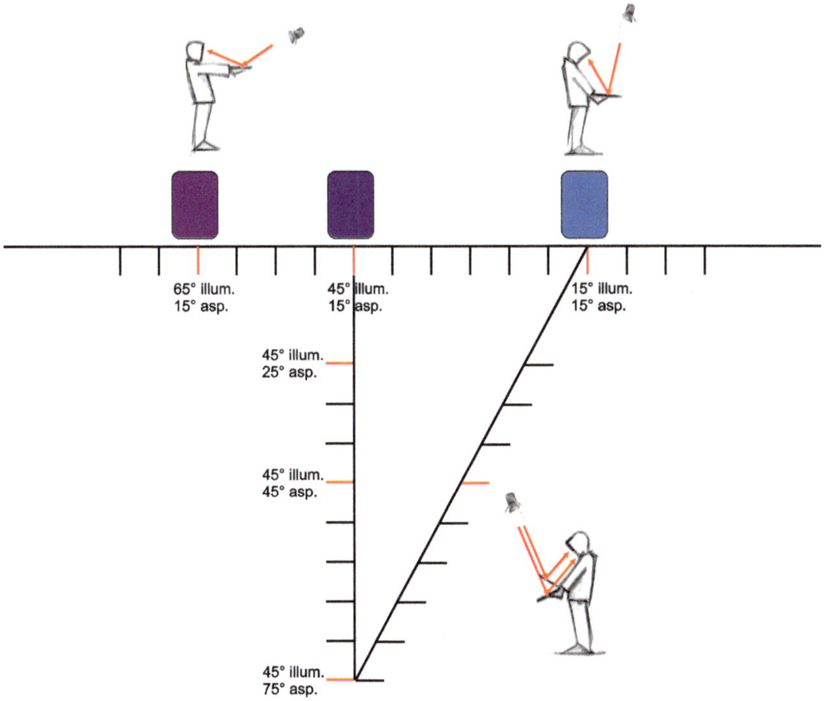

Figure 158: Schematic representation of the observations.

The geometries at the window, when a panel is tilted up and down, can be replicated using a laboratory instrument such as the Zeiss GK311/M. This provides, for the first time, an overview of the color travel in relation to the corresponding geometry. In this process, the angle between the light source and the observer remains constant.

Instrumental assessment

In the 1980s, interest grew in automotive paints with effect colors containing aluminum and interference pigments. In the mid-1980s, the first interference colors were introduced as white-series colors by Renault and Volvo.

At the same time, there was a growing desire for colorimetric assessment of these colors. Until then, colors had been measured either using directional geometry or spherical geometry. Both methods were and are insufficient for effect coatings because they do not capture and represent the effects. DuPont conducted visual tests with cans of various sizes that were coated with an aluminum pigment. These tests led to the angle set of 15°, 45°, and 110° from gloss at 45° illumination. For internal testing, DuPont built the DuPontmac with this angle set. At the same time, ZEISS selected the set with 25°, 45°, and 75° as ideal for displaying aluminum pigments.

Both sets were combined in the portable X-Rite MA-68, released in 1995, into one set with the five aspecular angles 15°, 25°, 45°, 75°, and 110° at an illumination angle of 45° (Figure 159). These geometries were randomly assembled and have not been verified. At the time of publication, there was little colorimetric experience with interference pigments. Later, this geometry set was adopted by other manufacturers without checking the necessity of these combinations.

In 1989, ZEISS began experimenting with a multiangle device in which the angles of illumination and observation/measurement could be variably adjusted. This led to the development of the GK311/M, in which an illumination head and an observation head are arranged on a steel half-rail. Figure 160 and Figure 161 show the GK311/M opened to the side with both heads in the geometric positions 70°:95° (20°:as15°) and 25°:140° (65°:as15°). Using software, they could be controlled independently of each other in 5° increments by motor. The minimum distance between the heads was 20°, and the flattest setting was 65° from the normal or 25° (protractor) from the horizontal.

My ideas, parts of the software, and the file format have all been incorporated into this device. I designed the latter in such a way that fast database queries are possible, and these measurement files can be opened with almost any program.

I have been using this device since 1991 to measure interference pigments and their mixtures, and in some cases with up to 252 geometries. This has enabled me to significantly expand my knowledge of the optical behavior of interference pigments. The optical properties of interference are reflected in measurements taken with steep, classic, and flat illumination at the same near-gloss difference angle. With the GK311/M, I was

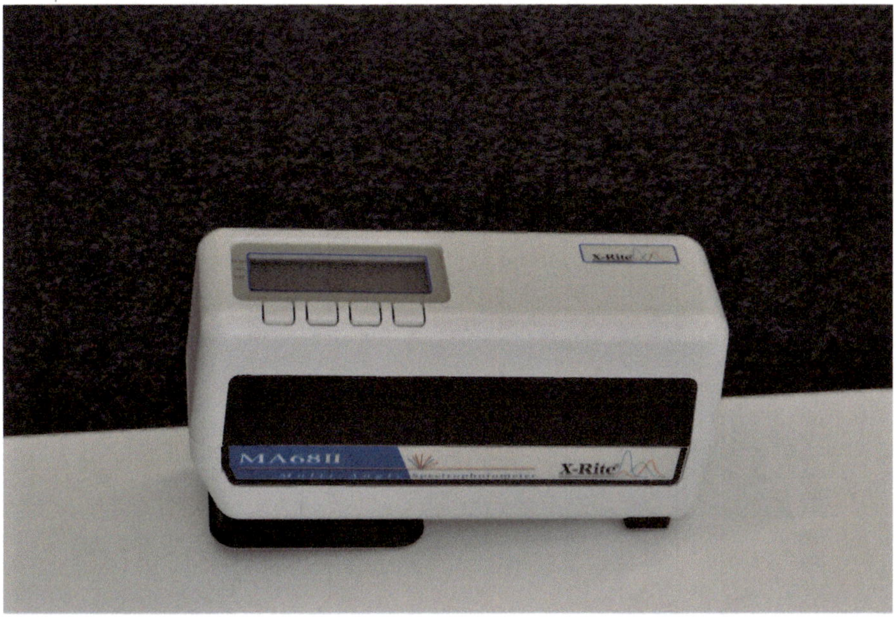

Figure 159: X-Rite MA68.

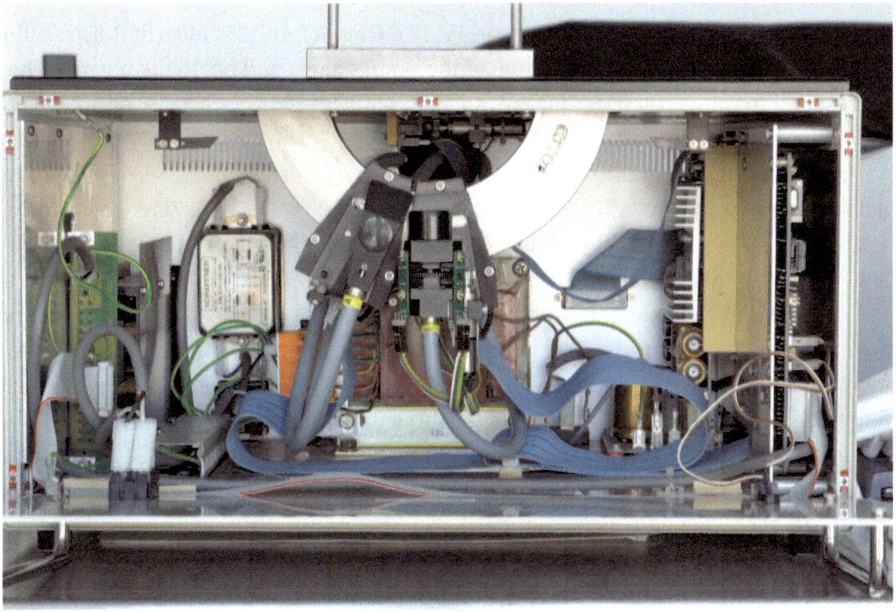

Figure 160: Zeiss GK311/M open side with 70°:as95° (protractor).

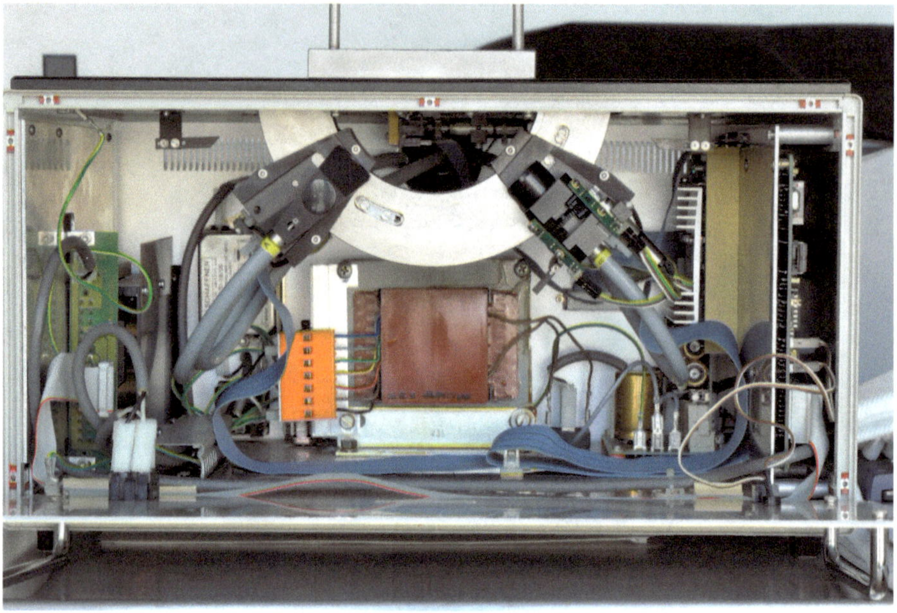

Figure 161: Zeiss GK311/M open side with 25°:as140° (protractor).

able to set the geometries 70°:95° (steep), 45°:120° (classic), and 25°:140° (flat). These illumination angles refer to the horizontal of the sample sheet and not to the normal. They correspond to the geometries 20°: –5°, 45°:-30°, and 65°:-50° – each 15° aspecular from the corresponding gloss angle.

As an alternative to the "normal" representation, in which the angles refer to the normal and are shown with plus/minus signs (e. g., +45° and –45°), Zeiss chose to use a semicircle above the panel with angles ranging from 0° to 180°. With an illumination angle of 45°, the gloss angle in this protractor display is 135°.

In 1999, the first meeting took place in Santa Rosa at Flex Products (now Viavi Solutions) to discuss interference pigments and their color measurement. I presented my findings here and in numerous meetings of the ASTM E12.12 committee, which was founded shortly thereafter, I advocated for three illumination angles, although flat illumination did not seem feasible for instrument manufacturers at the time. My ideas were then incorporated into Standard Method E2539 (then: Multiangle Color Measurement of Interference Pigments, now: Standard Test Method for Multiangle Color Measurement of Interference Flake Pigmented Materials), which was first published in 2008, with a second illumination at 15° and an aspecular angle of –15° for both illuminations (Figure 162).

Based on this standard method, BYK and X-Rite released their new measuring instruments: the BYKmac (Figure 163) has six measurement geometries at –15°, 15°, 25°, 45°, 75°, and 110° when illuminated at 45°. In addition, BYKmac offers the determination

Designation: E2539 – 25

Standard Test Method for Multiangle Color Measurement of Interference Flake Pigmented Materials[1]

This standard is issued under the fixed designation E2539; the number immediately following the designation indicates the year of original adoption or, in the case of revision, the year of last revision. A number in parentheses indicates the year of last reapproval. A superscript epsilon (ε) indicates an editorial change since the last revision or reapproval.

INTRODUCTION

Objects that exhibit a change in color with different angles of illumination and view are said to be "gonioapparent." The tristimulus colorimetric values of gonioapparent objects are derived using the spectral reflectance factors obtained from spectrometric measurements or colorimetric measurements at various angles of illumination and detection. The tristimulus colorimetric values are computed using the spectral reflectance factors of the object, the CIE Standard Observer, and the spectral power distribution of the illuminant, as described in Practice E308. This Test Method, E2539, specifies the color measurement of interference pigments at various illumination and detection angles.

Figure 162: ASTM E2539.

Figure 163: BYKmac.

of sparkle and graininess. For this purpose, the panel is illuminated at 15°, –45°, and –75° and photographed by a black and white camera at 0° in the normal direction. The photos are used to determine the values mentioned.

X-Rite published the MA-98 with the two recommended illumination angles and the difference angle at –15°, as described in E2539 (Figure 164). Off-plane measurements are also taken to capture the appearance of the pattern. In 2018, X-Rite launched the MA-T12 and MA-T6 (Figure 165). Both devices operate on the principle of light reversal, with illumination provided by 12 or 6 geometries for measurements at 45° or 15°. In addition, sparkle and graininess are determined using the 15° measurement.

To measure interference with three illumination geometries, the MA98 can use the two illumination angles of 15° and 45° with the respective difference angle of 15°. Due to the law of light reversal, 45°:as–15° is selected as the third geometry. This corresponds to a reversal of 60° illumination and a difference angle of +15°. In this respect, the measured values of 15°:as15°, 45°:as15° and 45°:as-15° = 60°:as+15° form the interference line (Figure 166, Figure 167).

The corresponding geometry selection is made in the MA-T12, which works with inverted light paths.

Since the BYKmac does not have a second light source, the geometries 45°:as15° and 45°:as–15° (= 60°:as15°) are used for identifying interference pigments. The connecting line of the color values a*b* between these two geometries always "bends" counterclockwise for colorful interference pigments compared to the line with the other aspecular

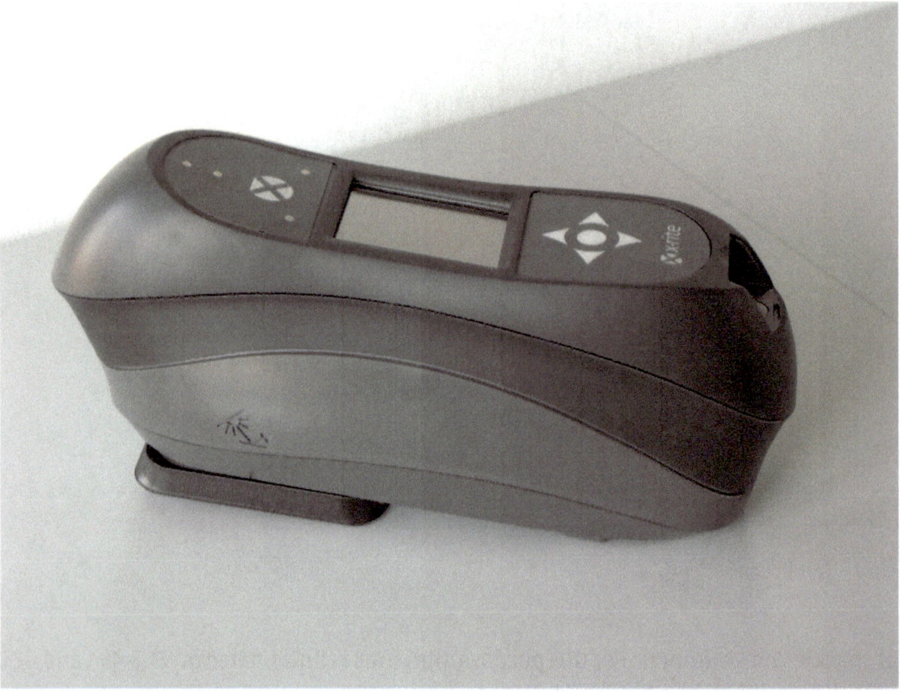

Figure 164: X-Rite MA98.

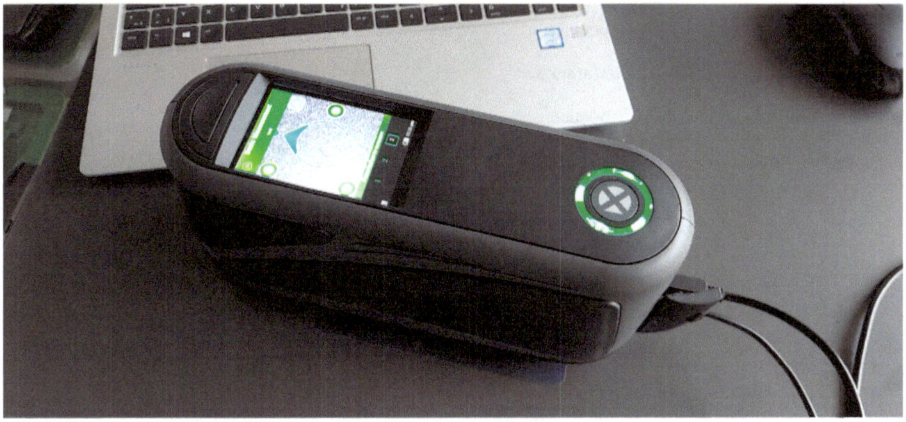

Figure 165: X-Rite MA-T12.

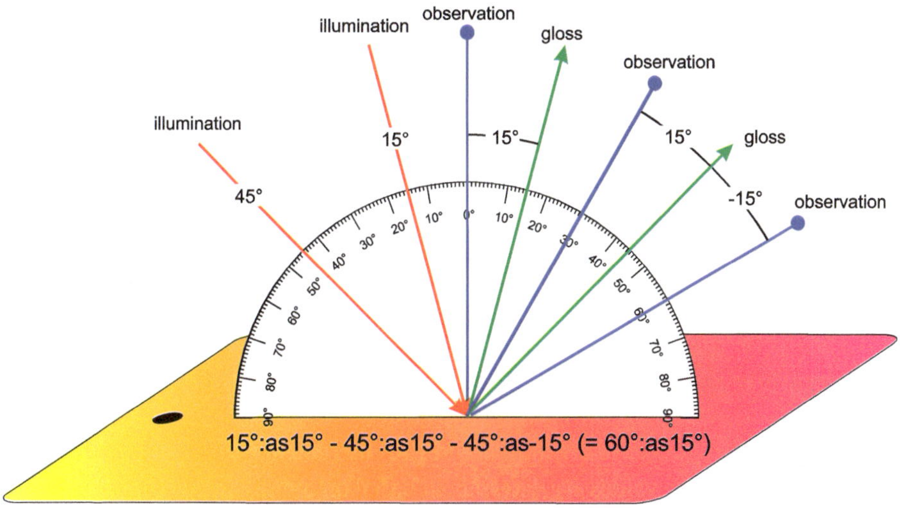

Figure 166: The interference line can be represented with the geometries 15°:as15°, 45°:as15° and 45°:as–15°.

color values. For aluminum pigments and white interference pigments, this line continues the line between the color values of 45°:as15° and 45°:as25°.

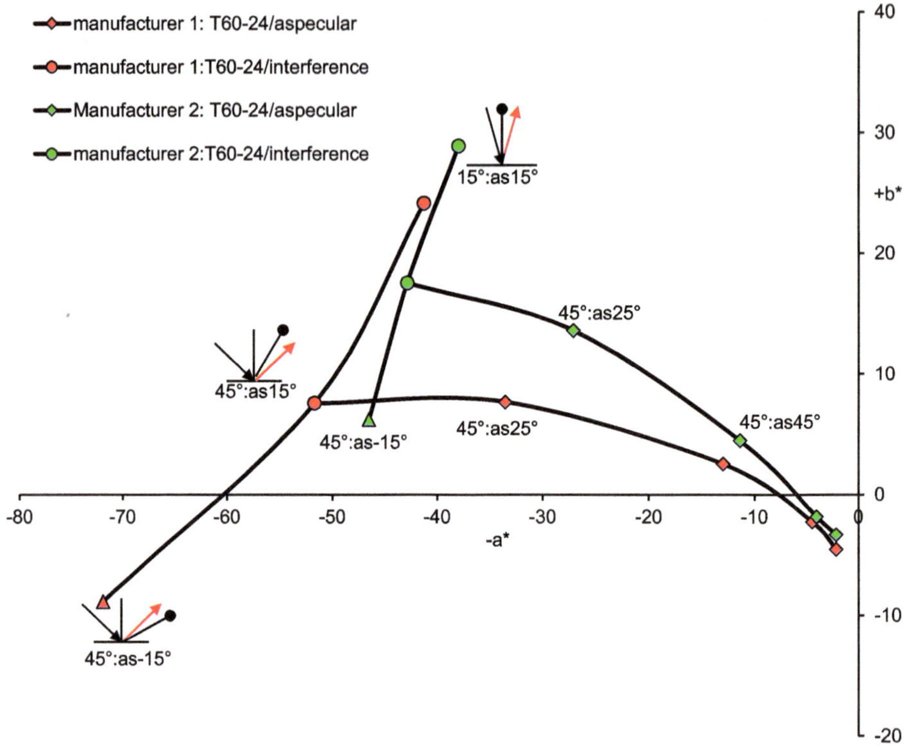

Figure 167: Different interference and specular lines from two manufacturers.

Reflectance values

Reflectance values are normally measured in 10° increments, sometimes in 20° increments. The reflectance values are determined based on the calibration of a device with a white standard. The white standard should be tested beforehand to ensure that it is also suitable for flat illumination angles. Tests with Teflon as a white standard did not produce plausible measurement results. White standards were measured against each other, as well as selected interference pigments. In Figure 168, the GK311/M measuring device was calibrated using the opal standard and the Spectralon white standard was measured. Significant differences are noticeable with flat illumination at 65°:-50°. Comparisons with calibration using Spectralon (Figure 169) and with opal (Figure 170) yield implausible measurement results for Stellar Green on Spectralon. Larger differences are found in measurements at –15° aspecular: Figure 171 leads to implausible results when calibrated with Spectralon, whereas calibration with Opal yields plausible results (Figure 172). The results show significant differences in the white standards, with Teflon falling behind in flat geometries (Figure 173).

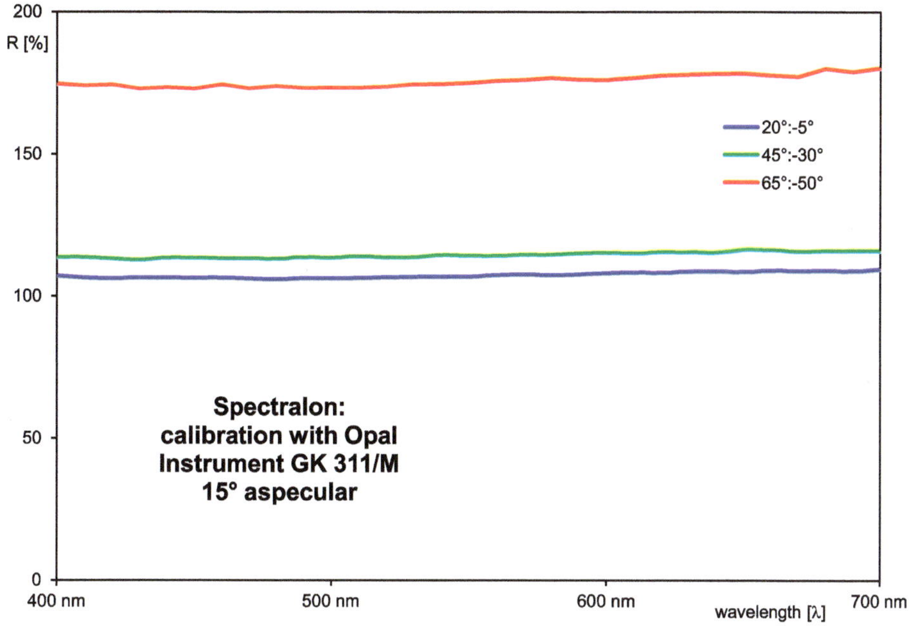

Figure 168: Spectralon measured, calibrated with Opal.

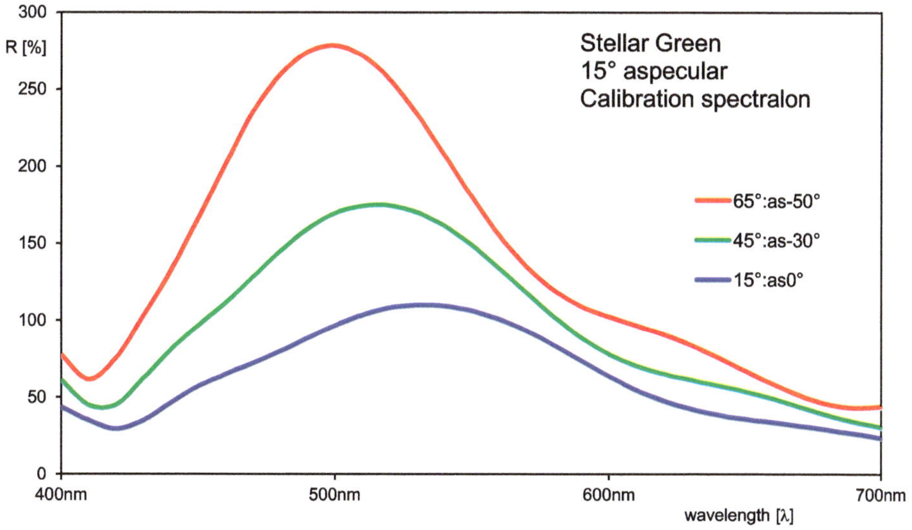

Figure 169: Reflection curves of Stellar Green at 65°, 45°, 15°:as15° calibrated with Spectralon.

Interference pigments, which normally show a shift in the reflectance curves toward shorter wavelengths and higher reflection in the shorter wavelength range, do not behave plausibly under flat illumination when calibrated with Spectralon.

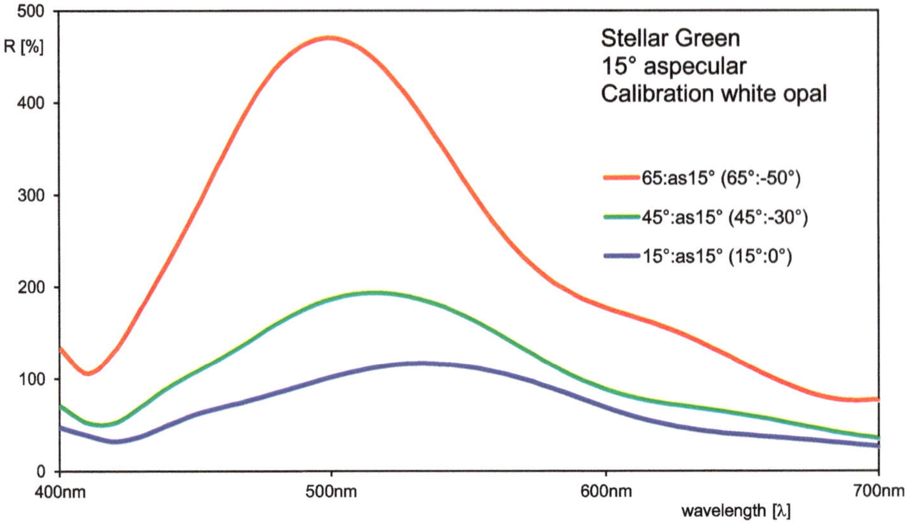

Figure 170: Reflection curves of Stellar Green at 65°, 45°, 15°:as15° calibrated with opal.

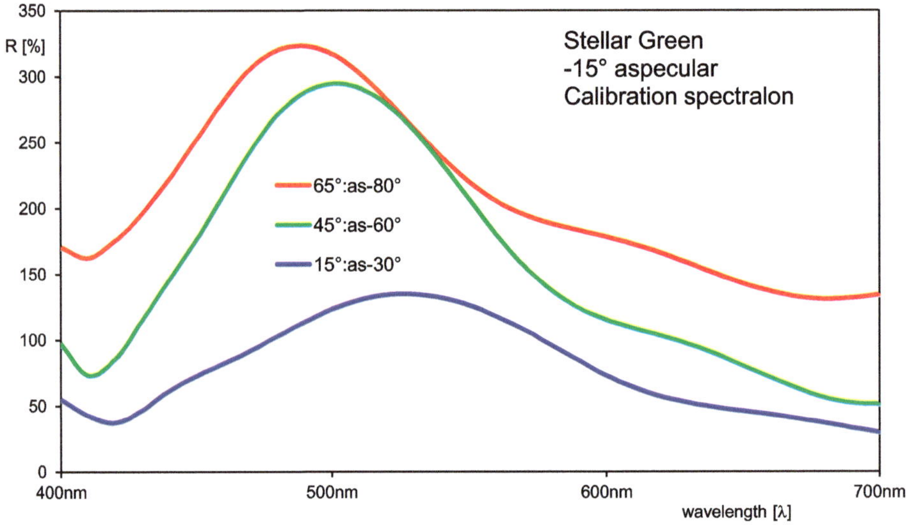

Figure 171: Reflection curves of Stellar Green at 65°, 45°, 15°:as−15° calibrated with Spectralon.

When describing interference pigments, their reflectance values for different measurement geometries are important because they directly reveal their physical and optical properties. They reflect these properties across the entire spectral range. As already mentioned, the reflectance curves of two interference pigments can be used to determine the nodes through which the reflectance curves of their mixtures pass.

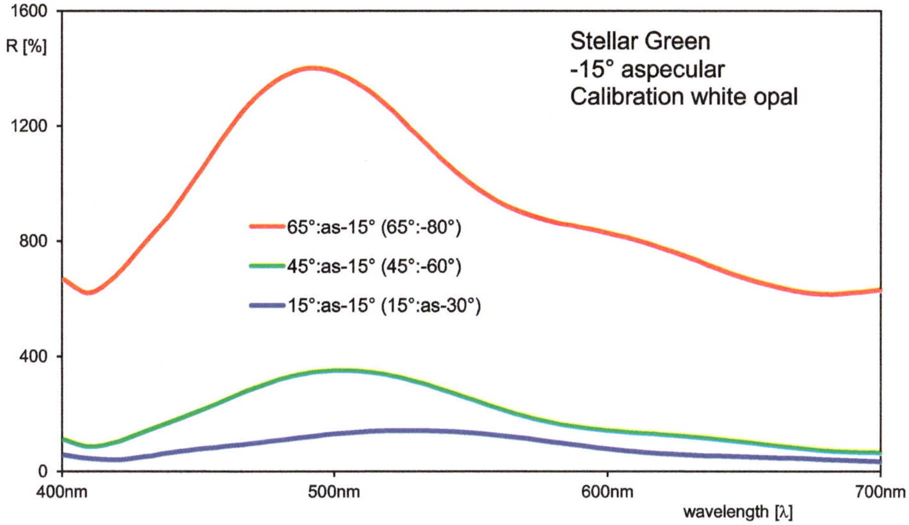

Figure 172: Reflection curves of Stellar Green at 65°, 45°, 15°:as–15° calibrated with opal.

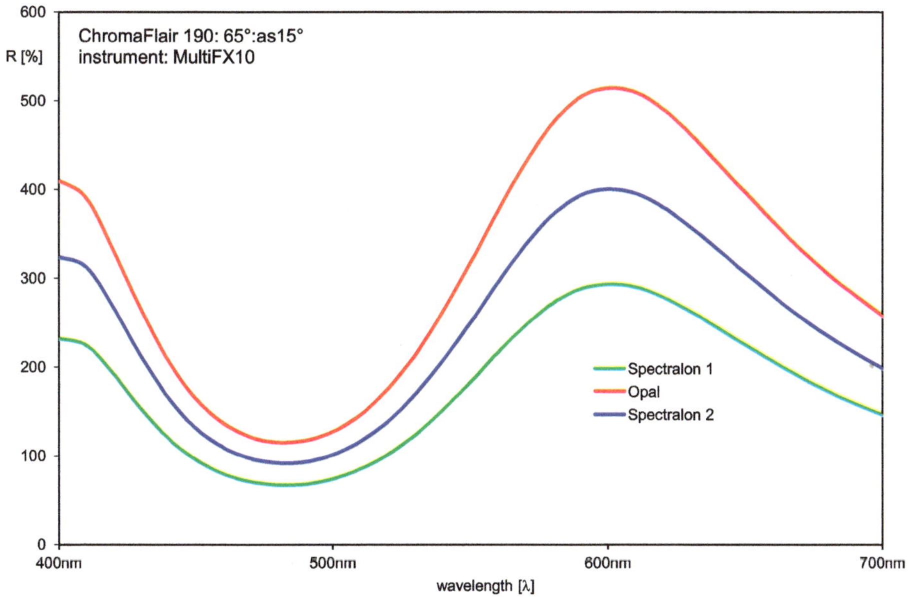

Figure 173: Reflections due to different white standards.

Number of measurement geometries

The three geometries are ideal for displaying the interference line: steep, classic, and flat illumination, for example, 15°, 45°, and 65°, each with a measurement and observation angle of 15°. ASTM E2539 lists two illuminations at 15° and 45° as the standard method. Here, the 45°:as–15° geometry can be used in reverse as 60°:as+15°. BYKmac has only one illumination and, with 45°:as+15° and 45°:as–15°, two geometries of the interference line.

The aspecular line begins with the same measuring point as the interference line at 45°:as15°. The next geometry, at 45°:as25°, lies in the transition area between reflection and transmission. For this reason, the measured values for this geometry are uncertain. 45°:as45° and 45°:as75° are stable geometries, while the following 45°:as110° often yields implausible results.

Calculation of color values

The physical reflection values can be converted into physiological color values. This corresponds roughly to the reaction of our brain to light rays that hit the retina of our eye, where they are converted into an optical stimulus, which in turn is "translated" into colors in the brain.

The International Commission on Illumination (CIE = Commission Internationale de l'Éclairage) had already considered linking the physical triggers of color stimuli with color perception in the 1920s. In 1931, the CIE standard color system, also known as the CIE standard color system, was published. This is based on experiments by David Wright and John Guild. With the help of so-called normal observers, colors were visually reproduced. They had to reproduce a specified color using the three light colors red (700 nm), green (546.1 nm), and blue (435.8 nm). A color was specified on one side of a divided screen, which the observers on the other side of the screen had to match by adjusting the three light sources. For certain specified colors, a match was only possible if red light was added to the mixture. This resulted in negative values for the red primary color (Figure 174).

The values of normal observers can be converted into positive spectral values:

$$\begin{aligned} x &= 2.7689r + 1.7518g + 1.1302b \\ y &= 1.0000r + 4.5907g + 0.0601b \\ z &= 0.0565g + 5.5943b \end{aligned}$$

They still apply today and are specified for both the 2° and 10° standard observers. These degrees refer to the respective field of vision, for example, with a field of vision of 2°, you are looking at a coin held in front of your eye with your arm outstretched. If you swap the coin for a postcard, the field of vision is 10°. These experiments established a

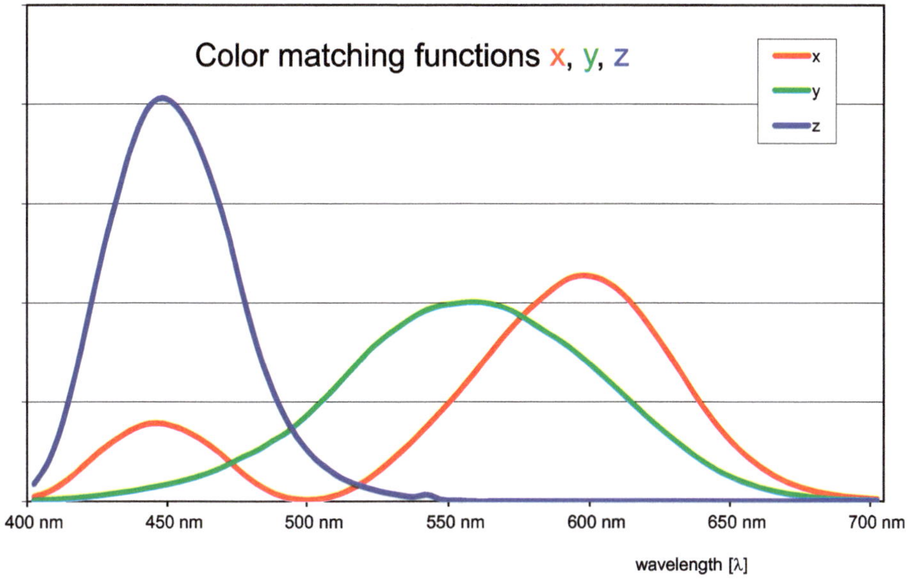

Figure 174: Color matching functions.

connection between the rays of light described by physics and the colors described by physiology.

For further calculation, the spectral power distribution of the light source is also required, for example, standardized daylight D65 (Daylight 6500° K, now D 6504° K) (Figure 175). It is multiplied wavelength by wavelength with the three spectral value functions. The calculated values yield the standard spectral values relative to the light source (Figure 176).

Calculating the area under these curves (= integral) yields the standard spectral values X_n, Y_n, and Z_n. Y_n is set to 100 because this value reflects the highest sensitivity of the eye. All values are converted to $Y_n = 100$. Further conversion yields the normalized spectral value components x, y, and z (not to be confused with the determined ones):

$$x = \frac{X}{X + Y + Z}$$
$$y = \frac{Y}{X + Y + Z}$$
$$z = \frac{Z}{X + Y + Z}$$
$$x + y + z = 1$$

If you plot the x and y values in a diagram – the z value is calculated using the formula – you get the two-dimensional chromaticity diagram (horseshoe shape) in the xy coordinate system (Figure 177) The outer line of this diagram forms the spectral color range.

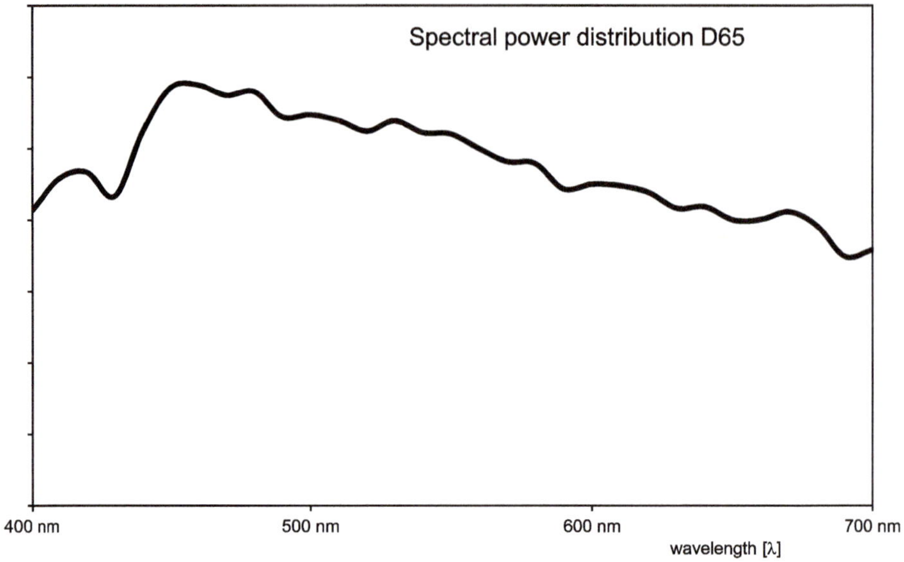

Figure 175: Spectral power distribution D65.

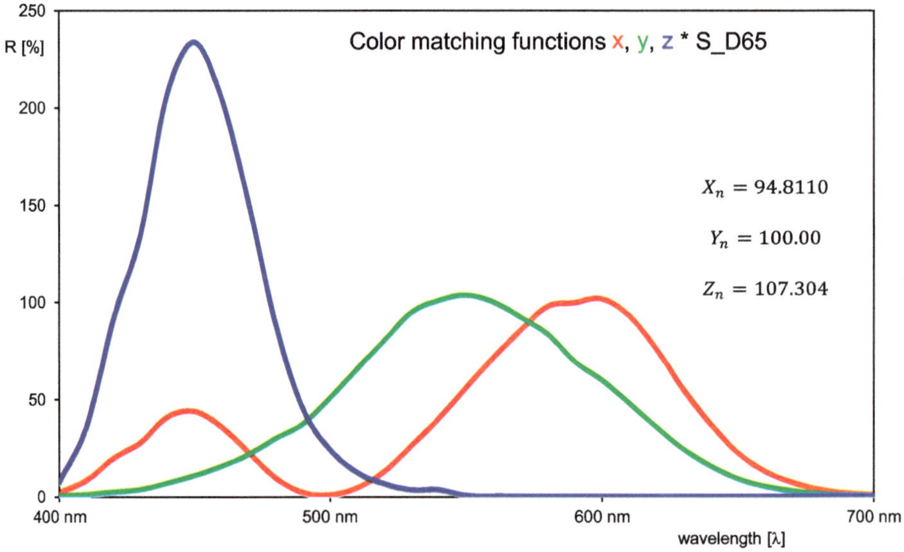

Figure 176: Color matching functions x Spectral power distribution.

And the ends of this line are connected by the purple line, on which all purple colors lie that do not occur in the visible spectrum and only exist in the brain.

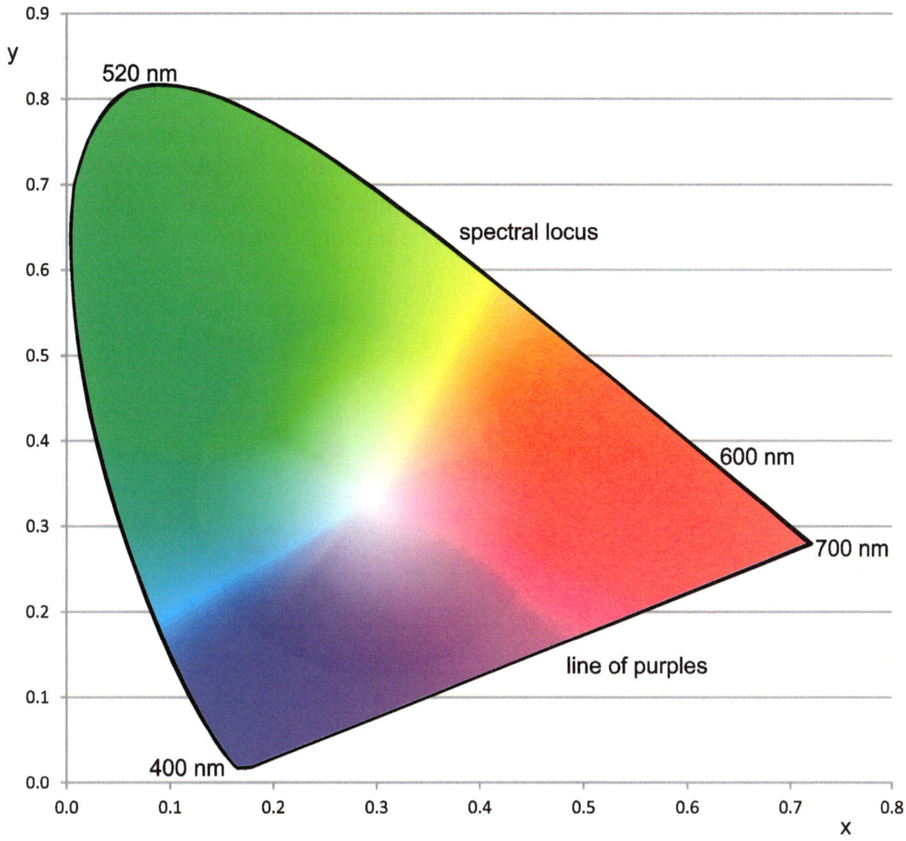

Figure 177: Chromaticity diagram.

The white point is in the center, for which the following applies:

$$x = y = z = 0.33$$

If you want to classify a color or color sample, it is measured physically: To do this, the sample is illuminated with white light from the measuring device and the reflected light is measured at equal intervals. The measuring instrument is calibrated beforehand with a white and a black standard. The measured values obtained are given as a percentage of the calibration white. These measured values are multiplied by the standard color values wavelength by wavelength. The area under the curves (= integral) provides the color values X_r, Y_r, and Z_r of the measured panel.

All perceptible colors can be classified in the standard color space. This standard color chart reflects the “translation” of light rays into physiological values. However, this type of representation is not very “handy,” which is why another step has been introduced. For this purpose, the X, Y, and Z values are converted into L*a*b* values. They then “fit” into a coordinate system in which each color has its exact position.

The coordinate system is based on a coordinate system with positive a* values for red and negative a* values for green. Positive b* values indicate the yellow component, negative b* values indicate the blue component. Each color is described by its red/green component (a*), its yellow/blue component (b*), and its brightness L* (Figure 178, Figure 179). These values are used every day in various industries around the world.

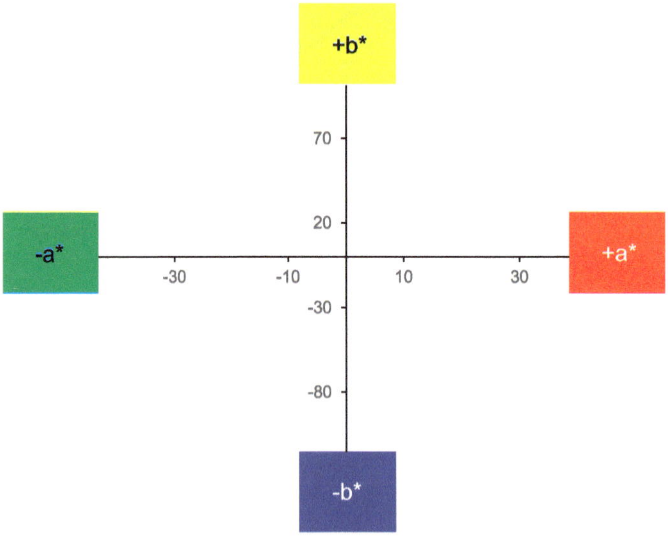

Figure 178: a*b* coordinate system.

Although the L*a*b* coordinate system is intended for calculating color differences, it can also be used very effectively for understanding color reactions.

Since different reflectance values and curves produce the same color perception, it is not possible to calculate reflectance curves or values from color values. The eye does not care whether the green – or the perception of green – comes from a single green pigment or a mixture of yellow and blue pigments.

In another representation, which can be calculated mathematically from the one presented, the color value is taken and its distance from the zero point is determined. This defines its chroma, i. e., its chroma C*: The further the color value is from the zero point, the more colorful the color (pigment) is. If you imagine a circle around the zero point, which starts at 0° and ends 360° counterclockwise on the red axis, you also obtain a value for the hue h°. The value of 0° corresponds to the red axis, 90° to the yellow axis, 180° to the green axis, and finally 270° to the blue axis. In addition to chroma C* and hue h°, the brightness L* is also required to uniquely describe a color (pigment). As shown in

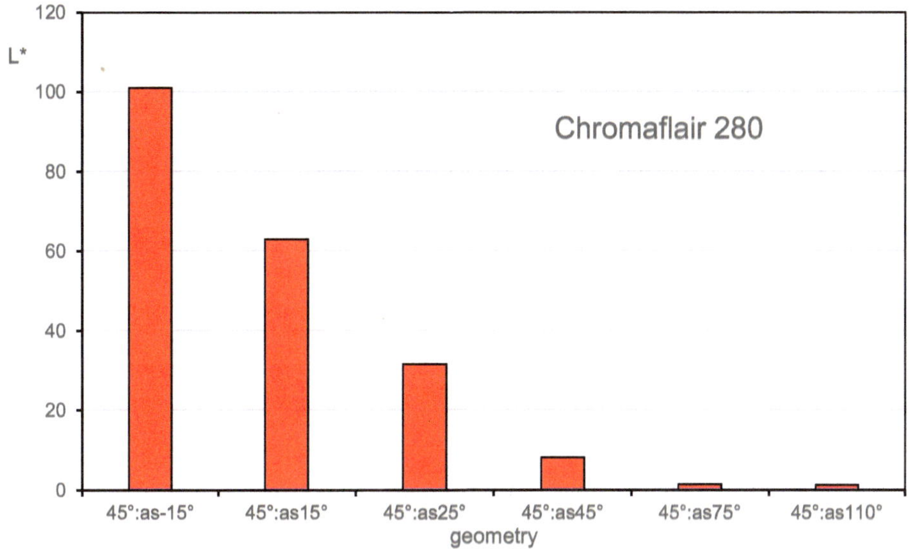

Figure 179: Brightness L* depending on the geometries.

the representation of the a*b* values, brightness L* is perpendicular to the coordinate system.

The representation of CIELAB values was originally used to calculate color differences. They are also suitable for describing interference colors and their color travel to understand their optical properties.

The software of the measuring instruments automatically calculates the CIELAB values from the measured reflection values. The display is usually based on individual geometries, for example, 45°:as–15°, 45°:as15°, 45°:as25°, 45°:as45°, 45°:as75°, and 45°:as110° (illumination:aspecular).

CIELab vs. OKLab

Both systems are based on XYZ color values. The OKLab system was introduced in 2020 by Björn Ottosson to better match the transitions between colors to our perception.

OKLab is implemented in CSS Color Level 4 (Cascading Style Sheets) and is supported by many web browsers. OKLab is more closely aligned with the perception of colors, particularly in the blue range. OKLab uses the same parameters as CIELab: L (brightness), a (red-green value), b (yellow-blue value). As with CIELab and its counterpart CIELCh, the OKLab system also has a representation known as OKLCh. Here, L represents brightness, C represents chroma (the distance from the origin), and h represents the hue angle starting from the red axis at 0°.

Both systems are color space systems that assist with color matching. Due to the very widespread global use of CIELAB, particularly in the pigment, paint, and automotive industries, the application of OKLab is concentrated in the web sector. When converting from RGB or CIELAB, the corresponding values are first converted back to XYZ. These XYZ values are then linearly converted to OKLAB values.

Like CIELAB values, OKLAB values are also device-independent.

6 Sparkle and flop

Wolf Moritz Cramer

In addition to the color of car paint, the surface structures (appearance) also became the focus of the analysis. These surface properties also include the so-called sparkle and flop of effect pigments.

Flop

In the 1980s, interest in metallic colors increased significantly, as did the registration of cars with these colors. At that time, paint manufacturers were looking for ways to capture and describe the effects of these metallic colors. Until then, colors had been measured using directional lighting and directional observation (45°:0° or 0°:45°) or spherical geometry. Both methods proved to be inadequate for metallic colors.

For this reason, DuPont developed a multiangle device called DuPontmac in the late 1980s. This device illuminated at 45° and measured at 15°, 45°, and 110° from the corresponding gloss angle at –45°. This combination was determined based on visual tests with cans and barrels painted with metallic paints (Figure 180).

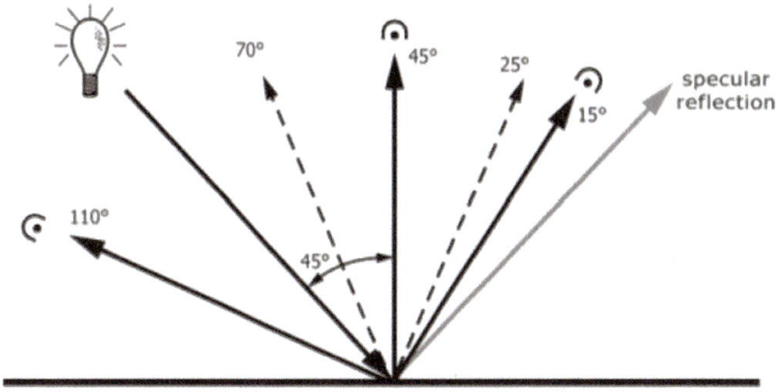

NOTE 1—Solid lines indicate preferred angles.

FIG. 2 Diagram of Aspecular Angles

TABLE 1 Preferred Angles

Uniplanar Angle	Preferred Angle
Near specular	15°
Mid-specular angle	45°
Far-specular angle	110°

NOTE 1—Other geometries in common usage are: 25°, 70°, or 75°.

Figure 180: Representation of the geometries of the DuPontmac.

https://doi.org/10.1515/9783111439204-006

David H. Altman used these geometries to establish his flop index (F_1), which subtracted the brightness at low gloss geometry from that at high gloss geometry and divided it by the brightness at medium gloss. The corresponding flop index was to be determined using appropriate exponents and factors.

This flop index was used not only for metallic paints with aluminum pigments, but also for paints with interference pigments. In principle, the formula makes sense for mathematically describing the flop based on the geometries. However, the flop index is of very little significance, as a flop is determined not only by an effect pigment, but also by the binders, additives, hardeners, thinners, and above all, the application method. If two painted panels are compared, the same flop index may have been achieved in very different ways.

Later, a flop index with modified values was also published. Instead of the angles close to and far from the gloss angle, observation angles of 25° and 75° from the gloss angle were used at the same illumination angle of 45° (F_2). This set was based on experiments conducted by the optics company Carl Zeiss (Figure 181).

$$F_1 = \frac{(L_{15} - L_{110})^{1.11}}{L_{45}^{\ 0.86}}$$

$$F_2 = \frac{(L_{25} - L_{75})^{1.11}}{L_{45}^{\ 0.86}}$$

Figure 181: Flop formulas according to DuPont (above) and Zeiss.

The flop indices refer to the brightness of a given panel. They refer not only to the pigments in the paint, but also to the application material and method. A flop is therefore influenced by various factors. All the following statements refer to the panels used and can only be used as guidelines for your own experiments. They do not allow any conclusions to be drawn about individual pigments.

Influence of the substrate color

Different substrate colors result in different flops for the same pigment. These values therefore do not refer to pigment, but to the combination of pigment and substrate color. It is therefore possible to influence the flop with the help of the substrate color.

The table lists the flop indices of the same pigment that was simultaneously coated over different substrate colors. In addition, the brightness values L* of the individual geometries and the chromaticity C* are listed (Figure 182).

The flop values for both formulas vary depending on the substrate color. The low values for the yellow substrate result from the high L* values for the near-gloss and

Crystal Silver	flop110	flop75	$L^*_{15°}$	$L^*_{25°}$	$L^*_{45°}$	$L^*_{75°}$	$L^*_{110°}$
gray	8.58	3.60	125.71	87.38	59.86	56.43	58.00
green	7.43	3.03	127.93	90.71	65.27	62.40	64.30
yellow	5.07	1.92	135.28	101.48	80.98	79.27	82.02
blue	8.08	3.35	126.46	88.61	62.09	58.74	60.49
red	7.71	3.14	127.25	89.58	63.83	60.78	62.59

Crystal Silver	flop110	flop75	$C^*_{15°}$	$C^*_{25°}$	$C^*_{45°}$	$C^*_{75°}$	$C^*_{110°}$
gray	8.58	3.60	6.36	3.04	0.20	0.16	0.28
green	7.43	3.03	12.97	17.83	28.07	31.17	32.92
yellow	5.07	1.92	7.45	20.06	34.41	37.28	38.56
blue	8.08	3.35	14.95	17.20	25.18	28.08	29.80
red	7.71	3.14	10.77	14.02	25.98	28.61	29.96

Figure 182: Measured results of Crystal Silver on various substrates.

far-gloss geometries. The difference is significantly smaller than for the other substrate colors (solid line = flop values) (Figure 183).

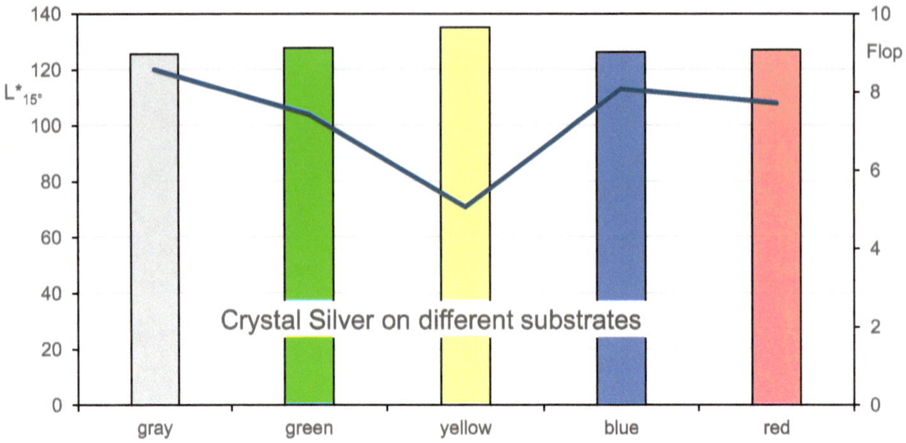

Figure 183: Same interference pigment, different substrate colors, different flops.

Comparison of black and white substrates

The influence of the substrate color is also evident when comparing Crystal Silver on black and white substrates (Figure 184). The flops are significantly lower on a white substrate than on a black substrate, which can be explained by the smaller difference in brightness between glossy and nonglossy areas. This difference is caused by reflections of the white substrate by the transparent pigment particles (Figure 185).

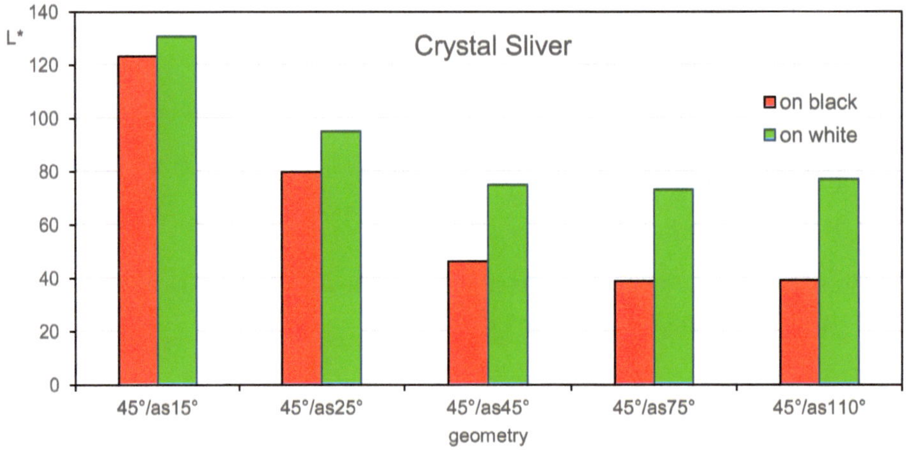

Figure 184: Crystal Silver on black and white.

Crystal Silver	Flop110	Flop75
on black	13.58	6.11
on white	5.45	2.03

Figure 185: Measured flop values.

Influence of particle size

CS = Crystal Silver, 9103 = Silverwhite, MS = MicroSilver (about 1/4 the size of CS)
Ratio Blue-green:CS/9103/MS from 0:100 to 100:0

With a lower concentration of the effect pigment, the flop increases. The lowest flop is seen in pigments without the addition of colored pigments. I created the panels using the ESTA Air process in a car manufacturer's laboratory, which is used in continuous coating at car manufacturers (Figure 186).

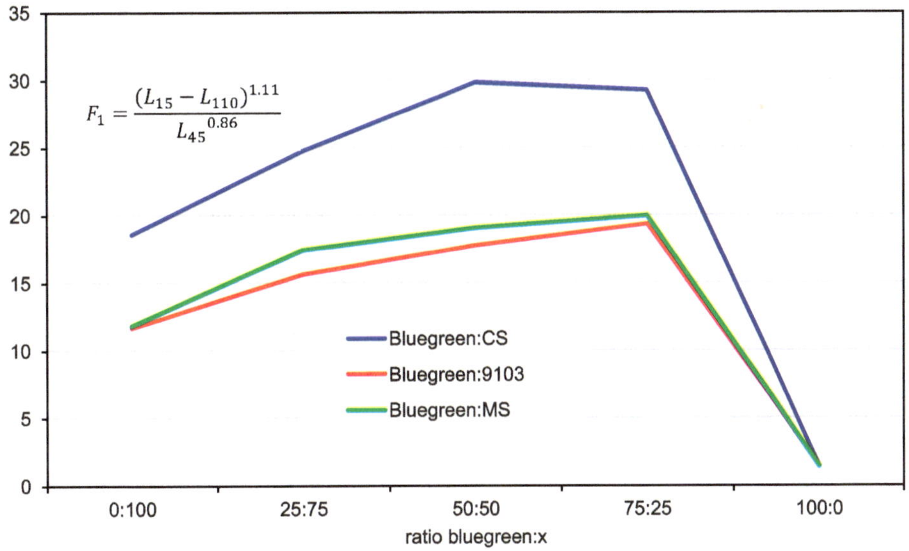

Figure 186: Flop of three interference pigments in blends with blue-green.

Influence of additives

Additives can be used to dramatically alter the flop in some cases. In this example, the brightness is reduced near the gloss and increased in the area away from the gloss.

The first diagram shows the brightness values observed when tilting the panels at the window or in a light booth (Figure 187).

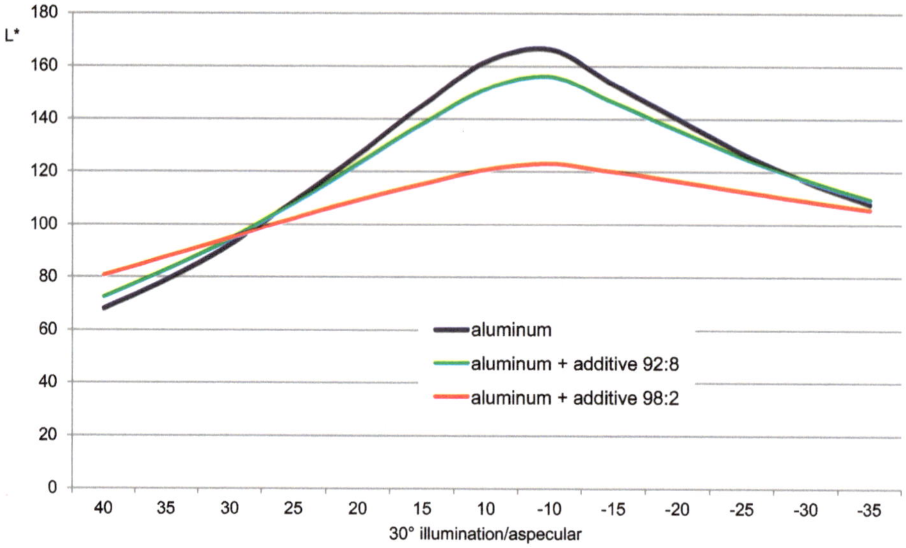

Figure 187: Brightness of the blends.

Measurements with the current geometries also show this behavior when additives are used: brightness is reduced close to gloss and increased in areas far from gloss. The result shows different flop values for the same pigment (Figure 188).

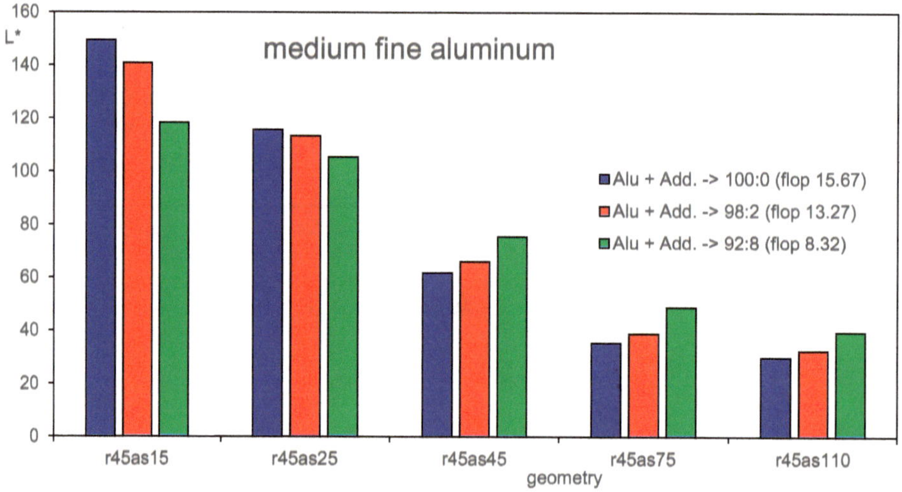

Figure 188: Blends with flop corrector.

Dependence on the spraying method

The same paint and pigment were sprayed differently. In the refinish sector, the application method can be changed by spraying the material faster, drier, or at a greater distance. The result can vary. In one experiment, a blue metallic paint was filled into a spray gun (refinish). Sixteen painters each sprayed a panel, resulting in 16 shades of the same paint (Figure 189).

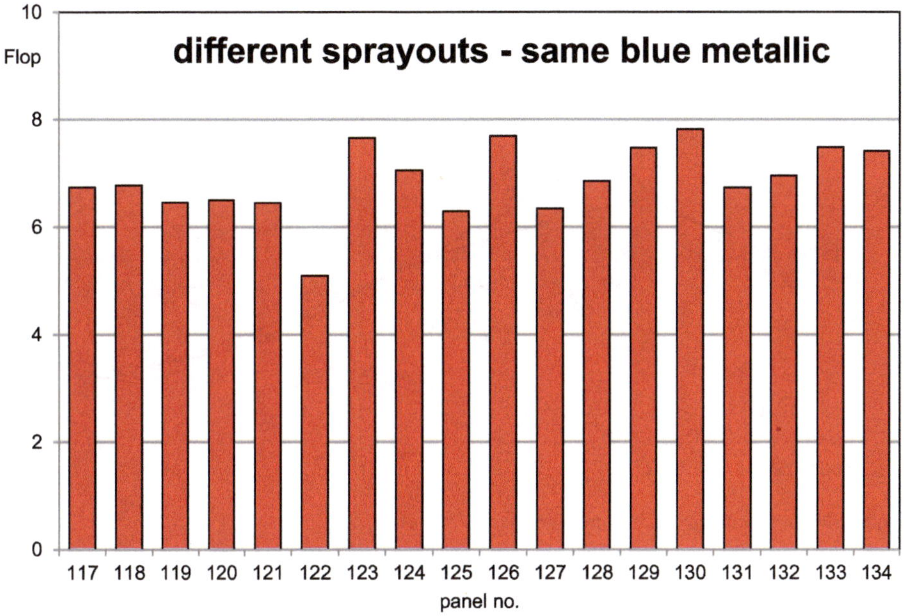

Figure 189: Sprayouts with the same material.

Different nuances, same flop?

Two delivered and accepted batches have different color shades but almost identical flop values. The user (car manufacturer) must decide whether to expect the same color or the same flop for both batches (Figure 190).

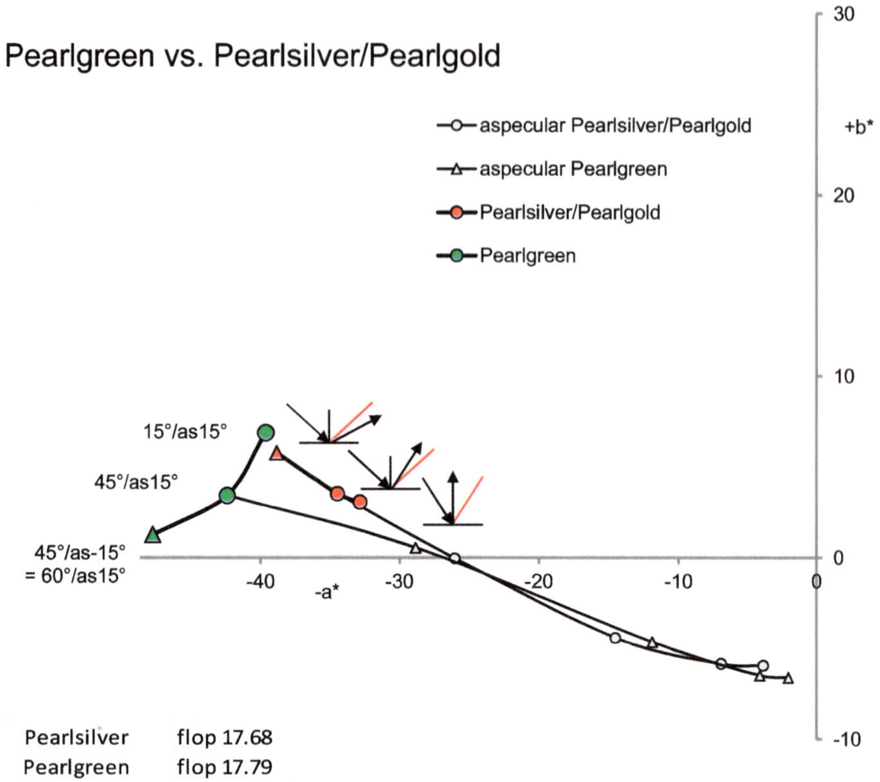

Figure 190: Different flops at Dragon Green.

Sparkle

When Merck introduced a new white interference pigment in 2000, car manufacturers' designers were enthusiastic about this "white metallic." However, it was not a "metallic" pigment, but a normal transparent interference pigment.

Because car manufacturers were unable to distinguish the new interference white from the familiar Pearl White in terms of colorimetry due to a lack of knowledge, they requested a description based on perceptions. This led to the development of the perception parameter "sparkle." The first studies were conducted as part of a joint project between BYK, AkzoNobel, and Merck under the name TARBAM.

BYKmac

First, BYK introduced the Sparkle parameter in the then-new BYKmac. It is calculated from black and white photos taken with a camera at the normal angle. The corresponding panel is illuminated at 15°, –45°, and –75° from the normal angle. On its website, BYK shows a graphical representation of the illumination angles at 15°, –45°, and –75° from the normal, whereas the specification states 15°, 45°, and 75°. This selection of angles is arbitrary and is based on the structural possibilities of the instrument.

In this respect, it is problematic to specify illumination angles of 15°, 45°, and 75° if these are not implemented in any other device. As mentioned above, the BYKmac illuminates at 15°, –45, and –75 from normal. If the measuring device is rotated 180° for different patterns, different sparkle values are obtained. For each illumination angle on the BYKmac, there are three parameters for the sparkle evaluation: S_i as sparkle intensity, S_a as sparkle area, and the total parameter S_G.

As with the statements on flop, many illustrations and publications reveal that panels with unknown pigmentations and application conditions were used for this purpose. Like flop, sparkle depends on the binder, the application, and the application conditions. In this respect, a sparkle value says nothing about the effect pigment. It does not refer to a specific pigment.

And panels are worthless unless the pigmentation and application are known. Car manufacturers are contractually prohibited from analyzing the paint. If, for example, an aluminum pigment is added to a paint with Crystal Silver, the sparkle behavior changes.

3nh: MS3012/MA-T12

To determine the sparkle, both devices use a 15° angle for measurement (principle of light reversal) and illuminate at difference angles of –45°, –30°, –15°, 15°, 45°, and 80°. These correspond to absolute angles of –60°, –45°, –30°, 0°, 30°, and 65°.

The sparkle value is now determined for many pigments, although it should be noted that aluminum pigments tend to glitter rather than sparkle. In addition, the interference pigment has the property that the sparkle impression increases when its concentration is *reduced.* With aluminum pigments, the glitter increases when the concentration is *increased.* In addition, Susonity's white interference pigment consists of two colorful interference pigments. If you compare the sparkle values of a white pigment with, for example, a blue or green interference pigment, you get different sparkle values, even though they are identical in structure.

Solids – i. e., blends without effect pigments – also deliver sparkle values, even though no effect pigments are present.

BRDF formula

Starting in 1999, mathematicians published articles on the topic of sparkle. They also made calculations based on BRDF. Later, these ideas were adopted by others. All considerations are theoretical in nature and have virtually nothing to do with reality. The BRDF formula does not consider the application material or the type of application (ESTA + Air, Air, etc.). Nor does it consider pigments, especially the optical properties of interference pigments. It also provides sparkle values when no effect pigment is present.

Experiments with sparkle

Visual assessment

To check the illumination angles of BYKmac, we built a GonioViewer 06 with five illumination angles on both sides of the normal: 15°, 30°, 45°, 60°, and 75°.

The BYKmac has illumination angles of 15°, –45°, and –75°. Since different sparkle values are determined for panels when the device is rotated 180°, the position of the illumination angles is important and decisive.

Visual experiments with the GonioViewer 06 show that the angle combination of 15°, 30°, and 45° is better than that of 15°, –45°, and –75°. Further experiments should therefore be conducted to confirm and verify the optimal combination (Figure 191).

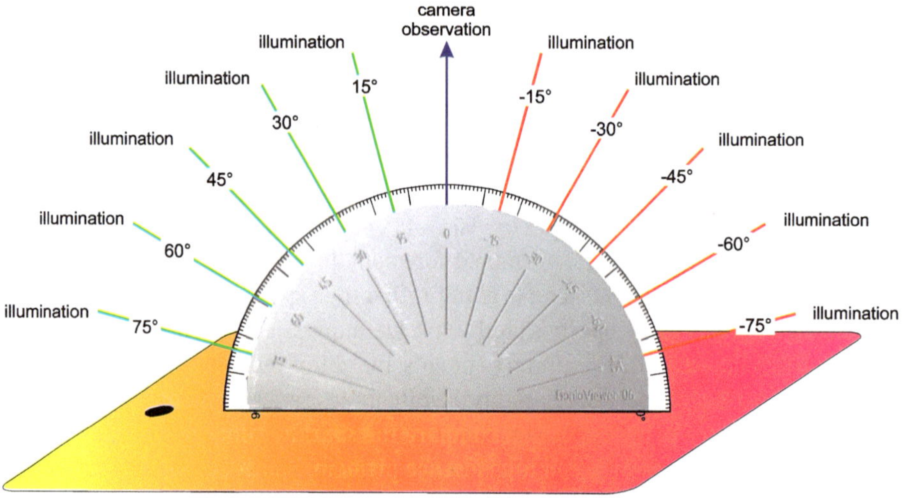

Figure 191: Sparkle geometries of the Gonioviewer 06.

Different colors, same sparkle

The determination – not measurement – of a sparkle value is not linked to the color of the panel. In this respect, different colors can have the same sparkle value. Furthermore, the sparkle value is not linked to a specific effect pigment. The specified sparkle value refers only to the present panel (Figure 192, Figure 193).

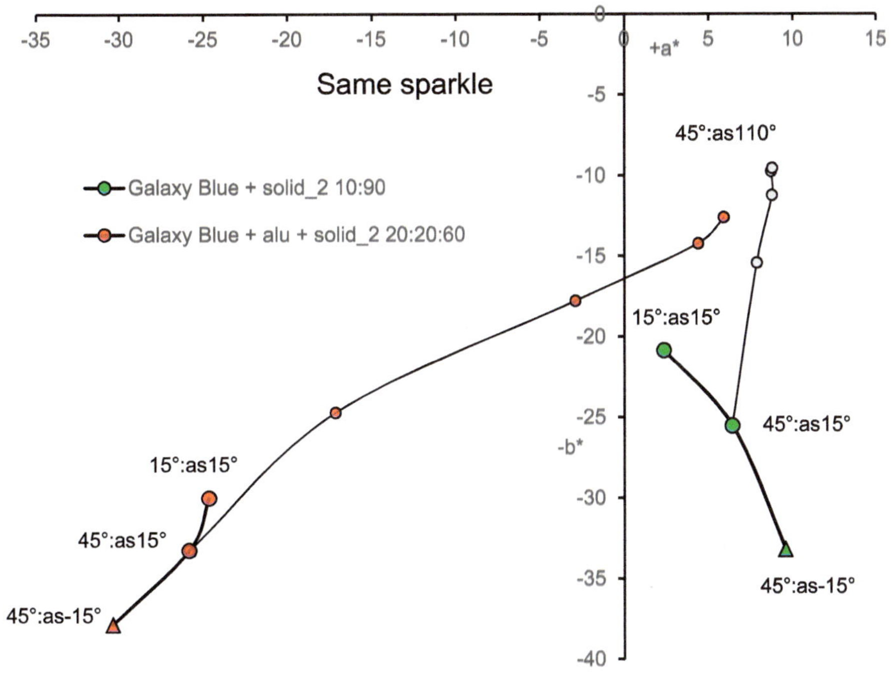

Figure 192: Same sparkle, different blends.

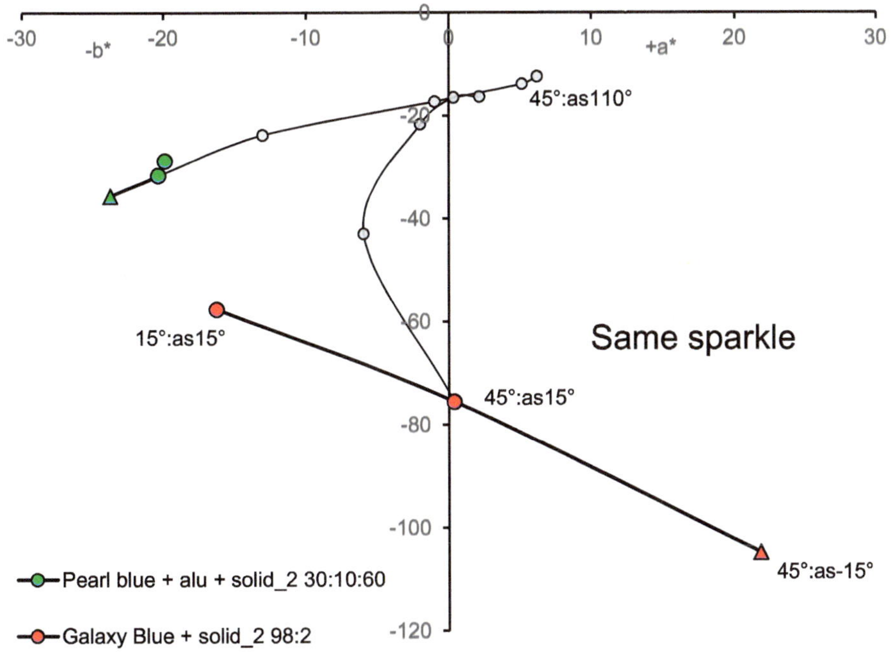

Figure 193: Same sparkle, different blends.

Solids

The measuring instruments and BRDF formulas also provide sparkle values for solids that do not contain effect pigments. Logically, these samples should not have any values. This demonstrates the inadequacy of sparkle determination (Figure 194).

Alignment

If the measuring devices are rotated by 180°, a series of panels will produce different sparkle values. This also applies to the formulas used to calculate sparkle (Figure 195).

Various application methods

The sparkle value changes depending on the application method (dry/wet, with and without effect layer (N1/N0)). However, it does not allow any conclusions to be drawn about the application method (Figure 196).

Dry or wet application

These methods are used particularly in repair painting. This allows the effect to be controlled and, if necessary, adapted to the repair part. With wet application, the effect particles have more time than with dry application to align themselves parallel to the surface (Figure 197).

Application with and without additional effect layer

For some automotive refinish coatings, it is recommended to cover the base coat with an additional effect coat after application. To do this, the air pressure at the spray gun or the spray speed can be increased. A larger spray distance also results in an uneven distribution of the effect particles. The example in Figure 198 shows a comparison of two sample sheets, whereby an additional effect layer was sprayed onto panel 9582 N1 compared to panel 9582 N0. This results in less sparkle in the three angles.

panel	date	channel	version	Coarsenes	Sparkle grade (brightness and contrast)	Sparkle grade (skewness)	Sparkle grade (spot)	Color parame	Color based sparkle grade (brightness and contrast)	Color based sparkle grade (skewness)	Color based sparkle grade (spot)	Sparkle grade (spot all channels)
green 3:1	2024_12_11	r15as15	Version 5.1	0.91	1.67	2.10	3.32	17.62	2.05	2.44	3.44	3.51
yellow	2024_12_11	r15as15	Version 5.1	0.42	0.04	1.22	1.04	4.33	0.06	0.81	0.72	1.22
red	2024_12_11	r15as15	Version 5.1	1.00	1.62	1.54	2.85	2.73	0.79	0.76	1.21	2.46
green 10:1	2024_12_11	r15as15	Version 5.1	0.94	1.29	2.03	2.69	15.31	1.58	2.22	2.74	2.76
white	2024_12_11	r15as15	Version 5.1	0.34	0.06	2.14	1.14	8.24	0.11	1.69	1.05	1.20
light gray	2024_12_11	r15as15	Version 5.1	0.19	0.14	1.94	1.00	7.67	0.21	1.52	0.92	1.17
dark gray	2024_12_11	r15as15	Version 5.1	0.42	0.86	2.09	2.56	15.48	1.17	2.28	2.65	2.51
black	2024_12_11	r15as15	Version 5.1	2.15	7.06	1.60	7.72	15.89	5.75	1.89	6.15	7.74
x1185	2024_12_11	r15as15	Version 5.1	0.25	0.00	0.05	0.00	1.97	0.00	0.05	0.00	0.39

Figure 194: Measurement results with solids.

		Flop	S_G 15	S_i 15	S_a 15	S_G 45	S_i 45	S_a 45
91040_north	Pearlblue/aluminum/solid_2	23.21	7.25	16.30	32.45	3.75	7.71	21.90
91040_west		22.83	6.94	15.26	32.05	3.51	6.82	22.27
91040_south		22.85	7.47	17.48	31.92	3.68	7.18	22.78
91040_east		23.10	7.02	15.15	32.96	3.23	5.20	25.56
91231_north	Galaxy Blue + solid_2	22.47	7.02	16.19	30.81	1.59	2.39	19.55
91231_west		22.97	7.11	16.39	31.19	1.58	2.91	15.84
91231_south		22.88	7.47	19.25	29.01	2.09	5.68	11.98
91231_east		22.88	6.90	15.22	31.80	1.89	3.55	16.67

Figure 195: Sparkle values of Pearl Blue and Galaxy Blue as a function of the measurement direction.

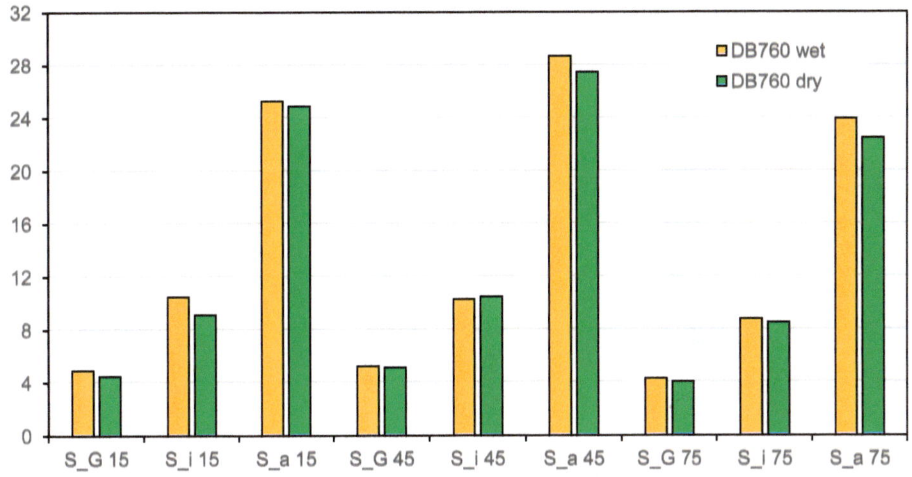

Figure 196: Car paint sprayed dry and wet.

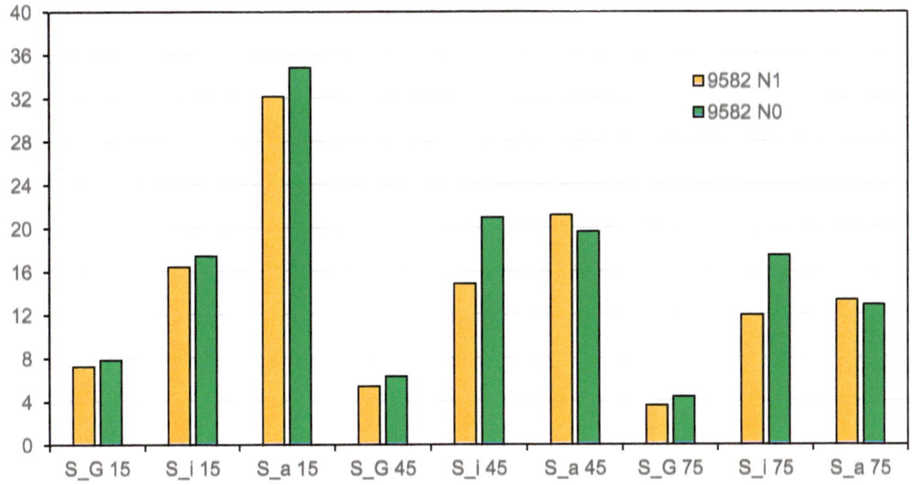

Figure 197: Interference paint with and without effect layer (N1, N0).

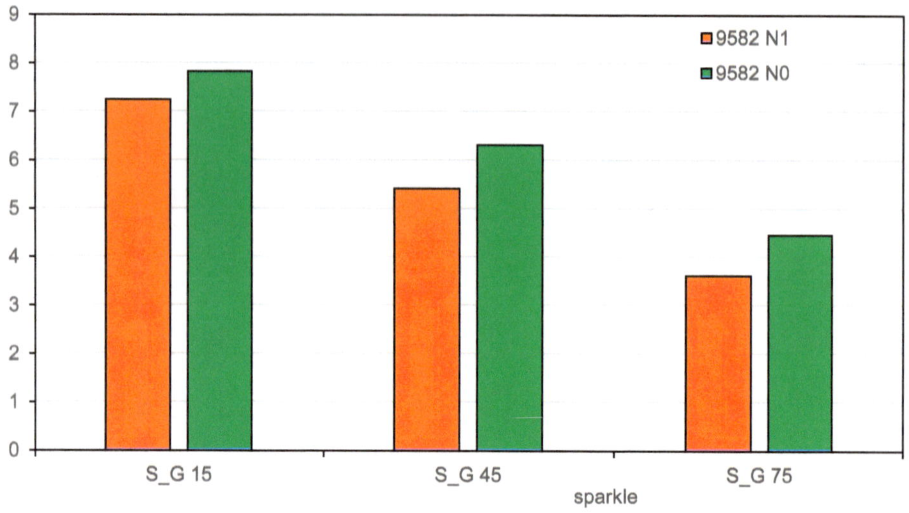

Figure 198: Interference paint with and without effect layer (N1, N0): S_G.

Replacement of effect pigments

In this panel with automotive paint, the white mica pigment (N) was replaced with Crystal Silver (X). When illuminated at a steep angle to determine the sparkle, the sample with Crystal Silver shows a significantly higher sparkle value (Figure 199).

In Figure 200, sparkle and flop values of the different pattern pairs are listed.

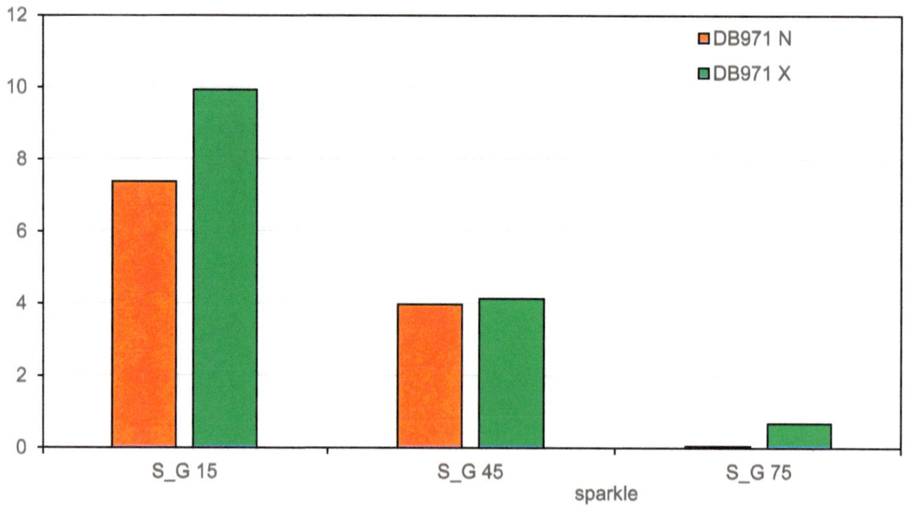

Figure 199: Interference paints with various interference pigments (N, X): S_G.

panel	Flop	S_G 15	S_i 15	S_a 15	S_G 45	S_i 45	S_a 45	S_G 75	S_i 75	S_a 75
DB760 dry	10.64	4.48	9.15	24.90	5.14	10.49	27.49	4.05	8.53	22.50
DB760 wet	9.21	4.90	10.51	25.27	5.22	10.30	28.69	4.28	8.81	23.93
DB971 N	11.94	7.39	14.79	37.05	3.99	7.30	25.59	0.05	1.15	5.15
DB971 X	14.55	9.94	44.20	21.32	4.14	10.61	18.76	0.69	1.96	9.24
9582 N1	15.84	7.25	16.45	32.18	5.42	14.88	21.22	3.63	11.99	13.35
9582 N0	18.71	7.84	17.48	34.84	6.32	20.99	19.72	4.46	17.51	12.91

Figure 200: Sparkle values of different sprayouts.

Tests at Audi

For a series of tests at Audi, the three interference pigments Crystal Silver, Micro Silver, and Pearl White were selected for base coats. Micro Silver is the smaller version of Crystal Silver in terms of flake size. All three pigments were added at a concentration of 5 % to a mixture with a green, blue, and black color pigment as base coats (Blue-green).

For the experiments, concentration mixtures were compiled, each containing the interference pigment in relation to the blue green in ratios of 100:0, 75:25, 50:50, 25:75, and 0:100.

When comparing the Sparkle value S_G 15 of the mixed series, the significantly higher values for Crystal Silver are striking (Figure 201). The values for Crystal Silver are also higher for sparkle intensity S_i 15. This demonstrates the behavior of Crystal Silver, whereby the sparkle increases with lower concentrations (Figure 202).

The sparkle area S_a 15 decreases significantly for Crystal Silver with increasing concentration compared to the other two interference pigments (Figure 203).

Looking at the brightness of the panels at a concentration of 75:25 (Blue-green: interference pigment), the values for Crystal Silver are lower than those for the other two interference pigments at all measurement angles.

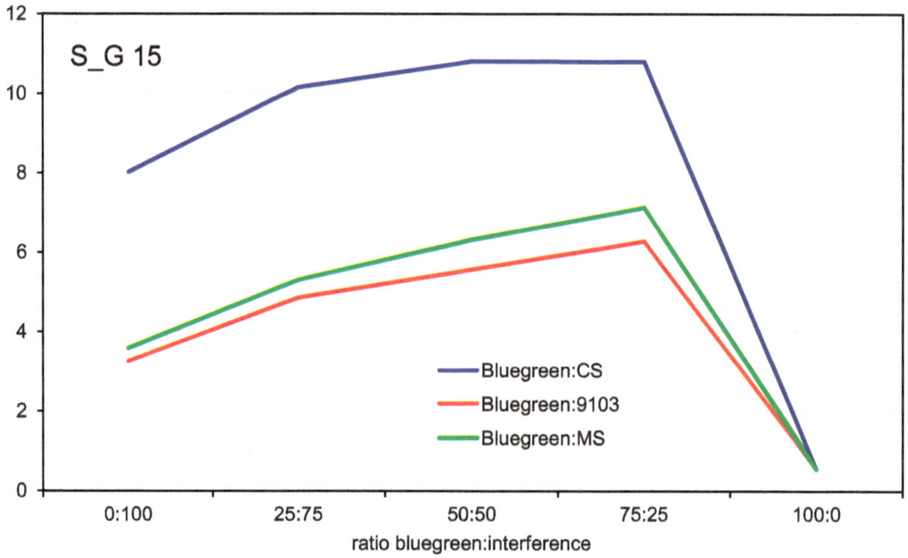

Figure 201: Sparkle determination with different interference pigments S_G 15.

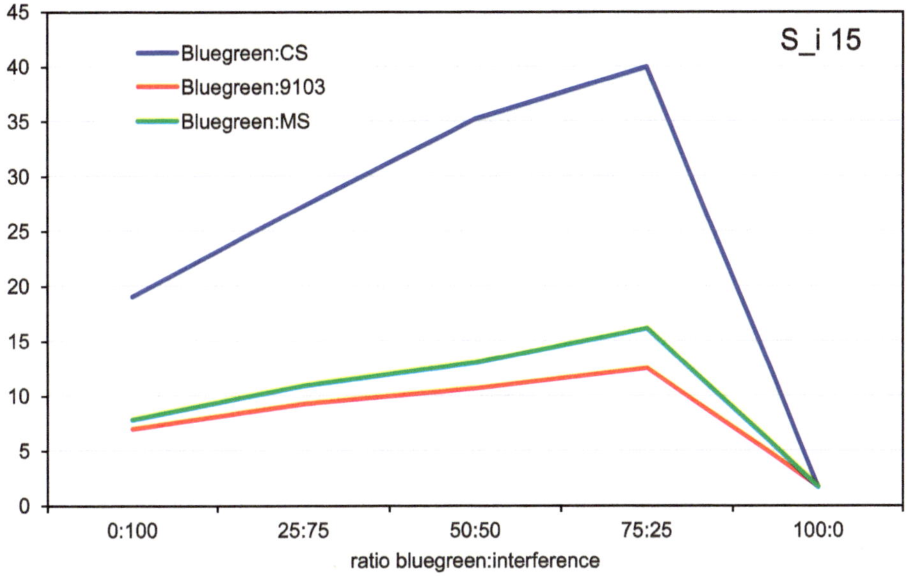

Figure 202: Sparkle determination with different interference pigments S_i 15.

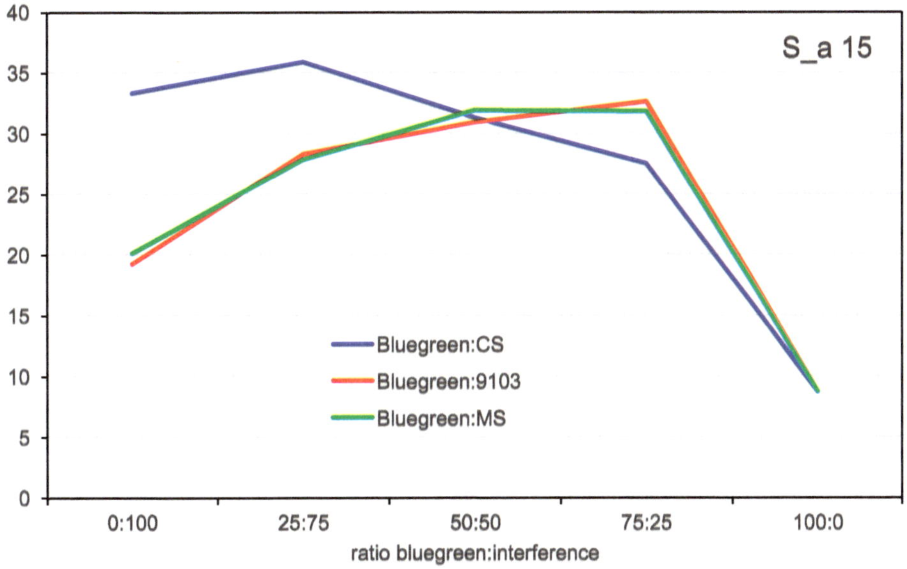

Figure 203: Sparkle determination with different interference pigments S_a 15.

Three dimensions of perception

At the heart of this approach to defining sparkle are three basic perceptual profiles that can be used to clearly characterize sparkle effects (Figure 204).

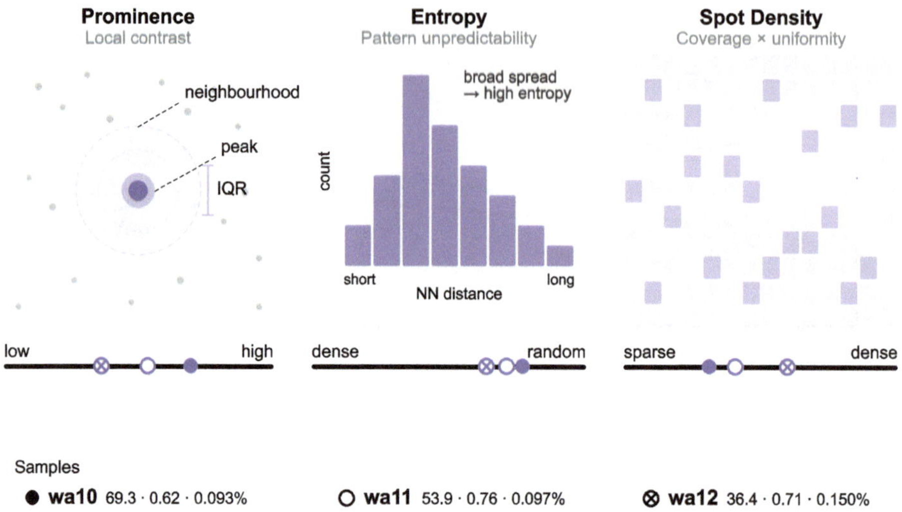

Figure 204: Presentation of prominence, entropy and spot density.

The first dimension describes the prominence of individual points of light. Sparkle only appears particularly prominent when it stands out visibly from its immediate surroundings. A bright reflection in a calm, darker environment is perceived differently than a similar reflection in a bright environment. This dimension captures the local prominence of a sparkle element and thus a significant part of what is experienced as "sparkle".

The second dimension concerns the unpredictability of the pattern. If the light points are very similar in spacing, the result is a more orderly and technical appearance. If these characteristics vary more widely, the surface appears livelier, more complex, and often more natural. This dimension of Entropy describes not the individual light points, but the character of the overall pattern.

The third dimension, Spot Density, describes how present the sparkle effect is across the surface – combining how much of the surface reflects light with how evenly those reflections are distributed. A surface with many, evenly scattered light points creates a different impression than one with few or strongly concentrated spots, even if the individual reflection appears equally prominent.

The three metrics are scale-independent.

7 Matte finish

Some standard colors are also offered by car manufacturers in a matte version (Figure 205). There are also suggestions for car repairs to paint parts in matte. The degree of matte plays a major role here, which refers to the clear coat and not the base coat.

Figure 205: Matte finishes.

Paint manufacturers offer paint systems in which the degree of mattness can be adjusted. The results of two series of tests are presented. Standox offers a matt paint that can be adjusted to the desired degree of mattness with a glossy clear coat. PPG offers a matt and a semimatt clear coat, both of which can be mixed to the desired degree of mattness.

Tests at Standox

Six sample sheets (postcard size) were sprayed simultaneously with the base coat Frozen Blue (BMW). Then clear coats were prepared with different mixing ratios of matt lacquer/gloss clear coat, different hardeners, and different thinners (Figure 206).

x0516	matte:gloss 70:30, Hardener 30 - 40, Thinner long
x0517	matte:gloss 75:25, Hardener 30 - 40, Thinner long
x0518	matte:gloss 80:20, Hardener 30 - 40, Thinner long
x0519	matte:gloss 75:25, Hardener 30 - 40, Thinner Express
x0520	matte:gloss 70:30, Hardener 10 - 20, Thinner long
x0521	matte:gloss 75:25, Hardener 10 - 20, Thinner Express
x0522	gloss CC, Hardener 30 - 40, Thinner long

Figure 206: Same base coat, different clear coats.

https://doi.org/10.1515/9783111439204-007

All results clearly show three characteristics of matte finishes.

First, they are brighter than their glossy variants, which can be observed both in the brightness values L* and in the reflection curves (Figure 207, Figure 208).

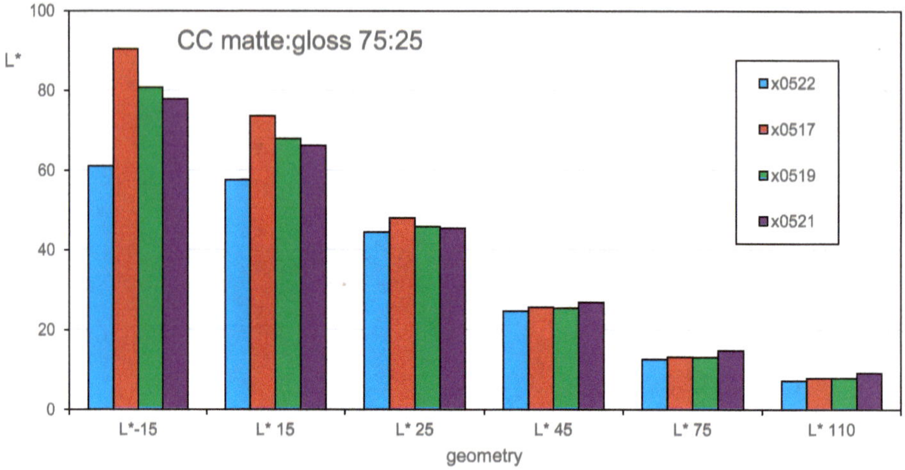

Figure 207: Same base coat, different clear coats.

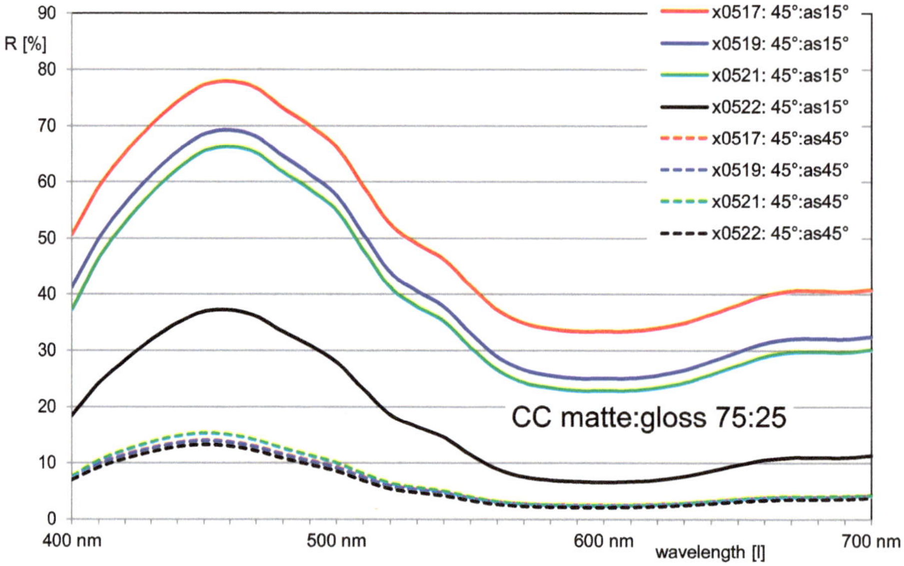

Figure 208: Reflection curves of the blends.

Second, their chroma C* increases from geometries close to gloss to geometry 45°:as25°, only to decrease again for geometries further away from gloss (Figure 209, Figure 210).

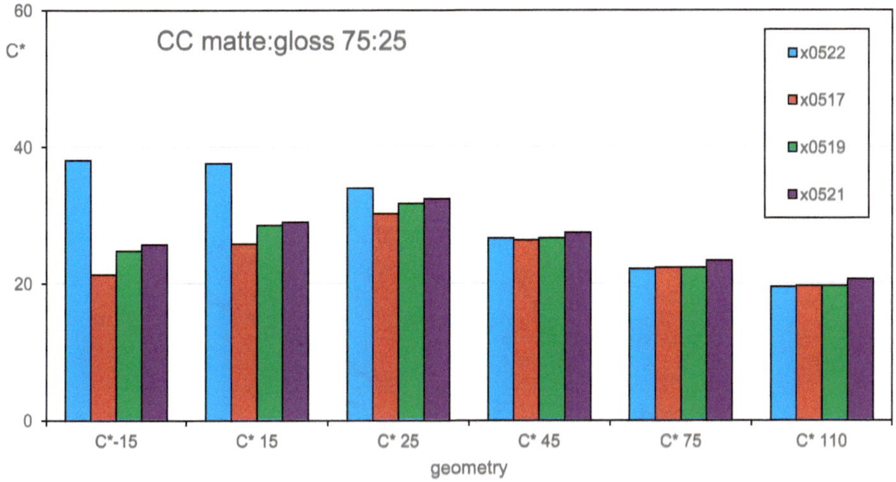

Figure 209: Chroma C* of the blends 75:25.

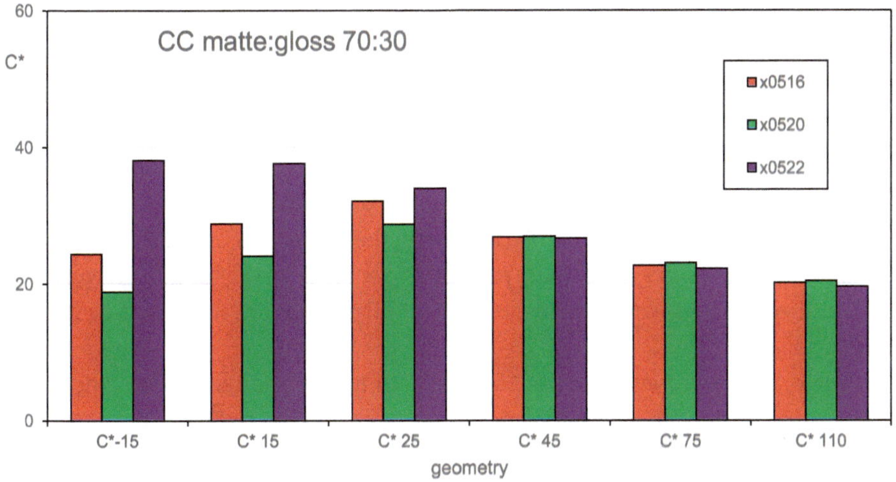

Figure 210: Chroma C* of the blends 70:30.

Third, the a*b* lines "turn" from 45°:as45° clockwise to the geometries close to gloss (Figure 211).

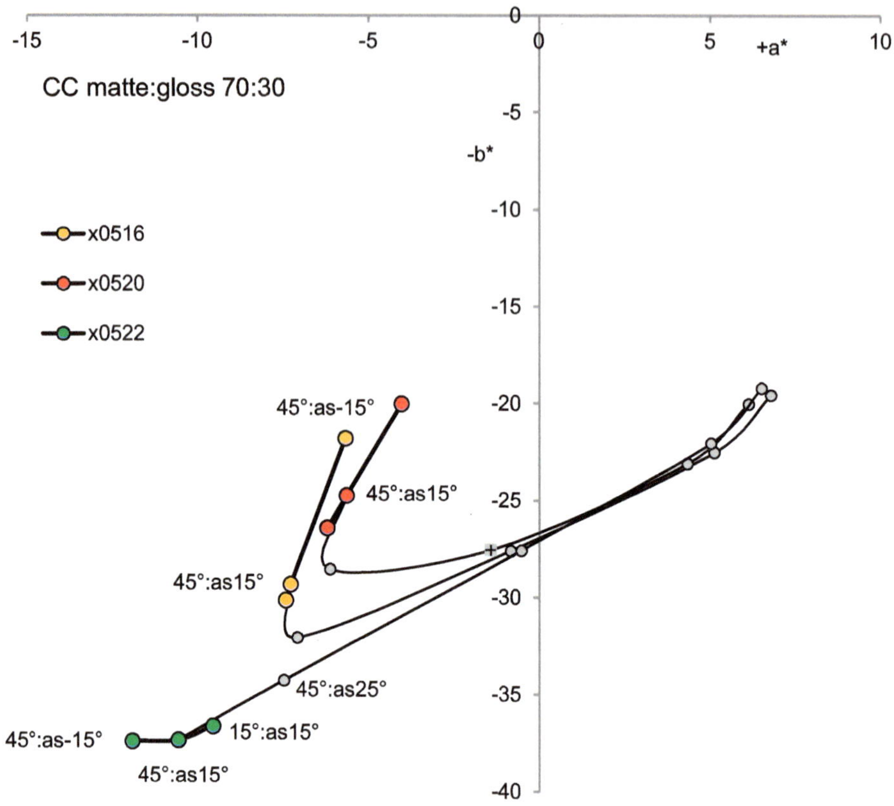

Figure 211: The interference line "bends" clockwise.

Tests at PPG

For these test series, we combined the three effect pigments Stellar Green, Galaxy Blue, and an aluminum pigment. These were added to the binder as a base coat (Figure 212).

By mixing two matt clear coats, you can adjust the degree of matting (Figure 213). The result of applying a matte clear coat is highly dependent on factors such as flash-off time, drying time, etc.

The reflection curves of Galaxy Blue also shift toward shorter wavelengths for the matte panel x604 when the illumination angle is flatter (Figure 214). If the chroma C* is plotted, it increases to geometry 45°:as25° for the panels of Galaxy Blue and Stellar Green (Figure 215, Figure 216). The brightness values are greater than those of the glossy reference panel (Figure 217, Figure 218).

The interference lines in the a*b* chart shift significantly clockwise in the case of matt-coated samples. Glossy samples, on the other hand, show a color shift counterclockwise.

		semi-gloss (D8115) (percentage)	matte (D8117) (percentage)
x0595	Stellar Green + CC (gloss)	0	0
x0600	Stellar Green + FC05	0	100
x0599	Stellar Green + FC05	30	70
x0598	Stellar Green + FC05	50	50
x0597	Stellar Green + FC05	70	30
x0596	Stellar Green + FC05	100	0

		semi-gloss (D8115) (percentage)	matte (D8117) (percentage)
x0601	Galaxy Blue + CC (gloss)	0	0
x0606	Galaxy Blue + FC05	0	100
x0605	Galaxy Blue + FC05	30	70
x0604	Galaxy Blue + FC05	50	50
x0603	Galaxy Blue + FC05	70	30
x0602	Galaxy Blue + FC05	100	0

		semi-gloss (D8115) (percentage)	matte (D8117) (percentage)
x0607	Aluminum + CC (gloss)	0	0
x0612	Aluminum + FC05	0	100
x0611	Aluminum + FC05	30	70
x0610	Aluminum + FC05	50	50
x0609	Aluminum + FC05	70	30
x0608	Aluminum + FC05	100	0

Figure 212: Blends PPG.

Gloss Level	FC01	FC02	FC03	FC04	FC05
	Matt e.g. Lamborghini full body Matt Finish		**Low Gloss** e.g. Mercedes, Smart, BMW, Fiat full body Matt finishes		**Semi Gloss** e.g. Older Mercedes Plastic side mouldings
Clearcoat	Percentage Parts by Weight (%)				
D8115	100	70	50	30	0
D8117	0	30	50	70	100

Figure 213: Compilation of clear coats.

Matte or semimatte clear coats change the color of an interference pigment. The chroma also increases initially when measured further away from specular. The bright-

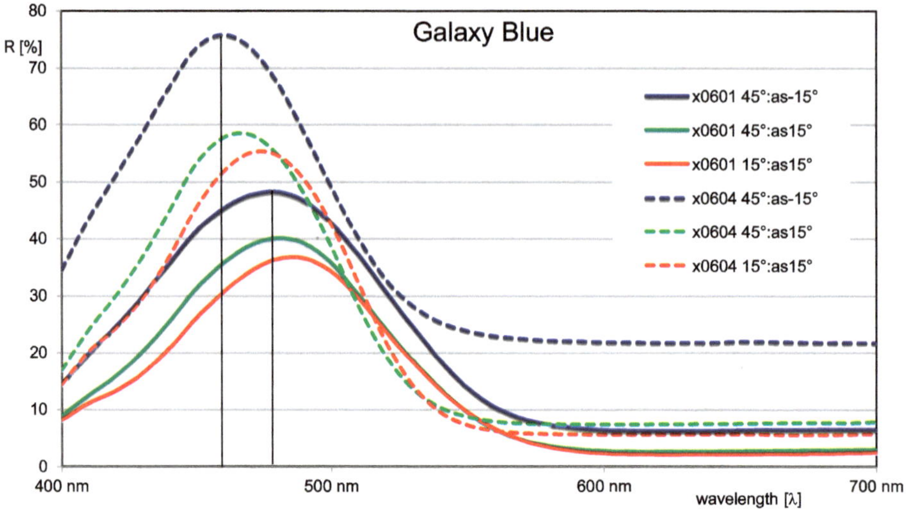

Figure 214: Reflection curves of Galaxy Blue.

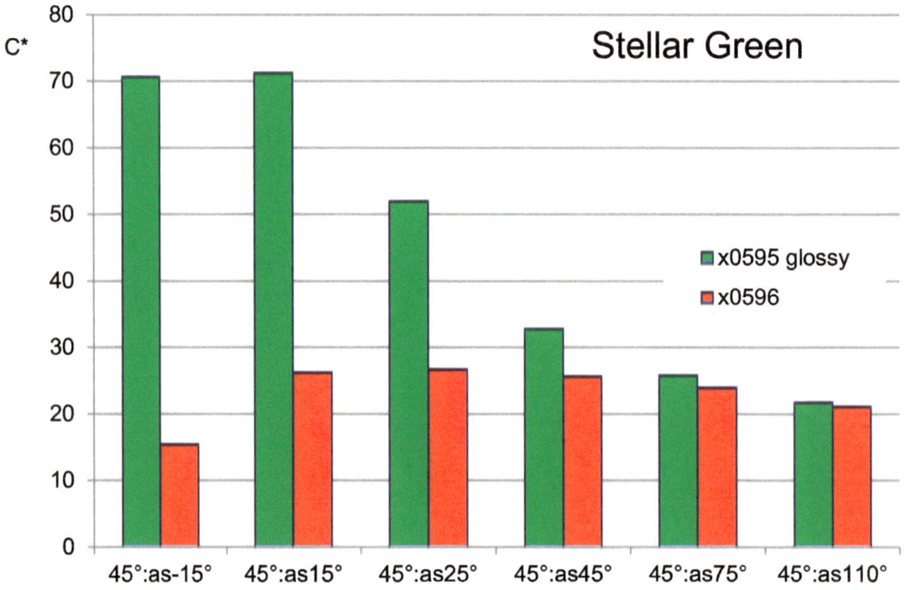

Figure 215: Chroma C* Stellar Green.

ness values L* are significantly higher for matte coatings than for glossy coatings. Visually, matte coatings appear milky, which indicates a higher degree of scattering.

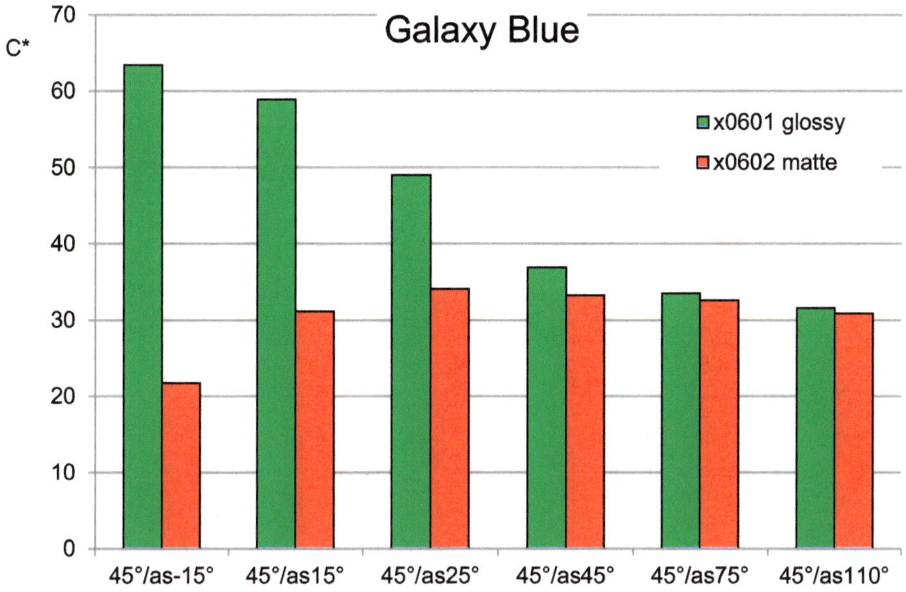

Figure 216: Chroma C* Galaxy Blue.

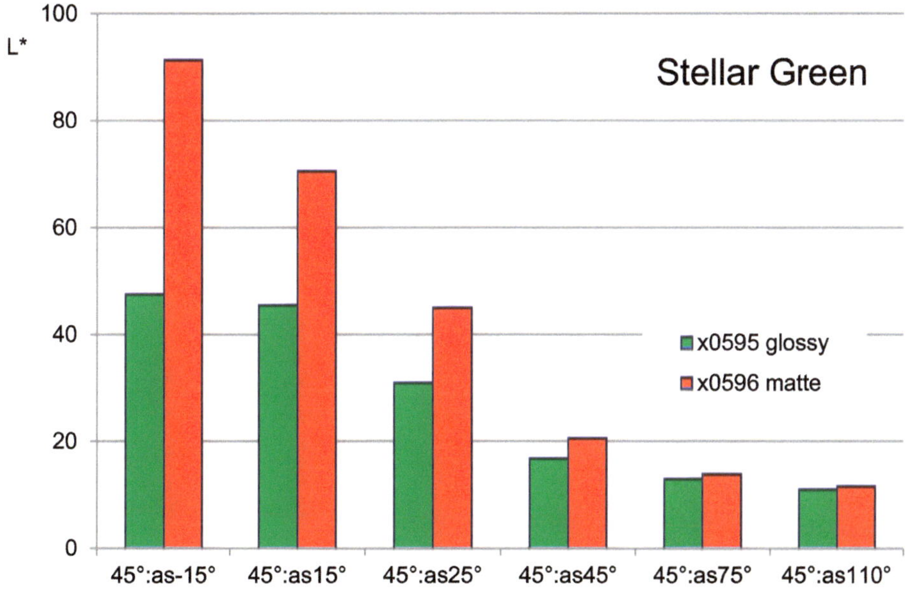

Figure 217: Brightnesses L* Stellar Green.

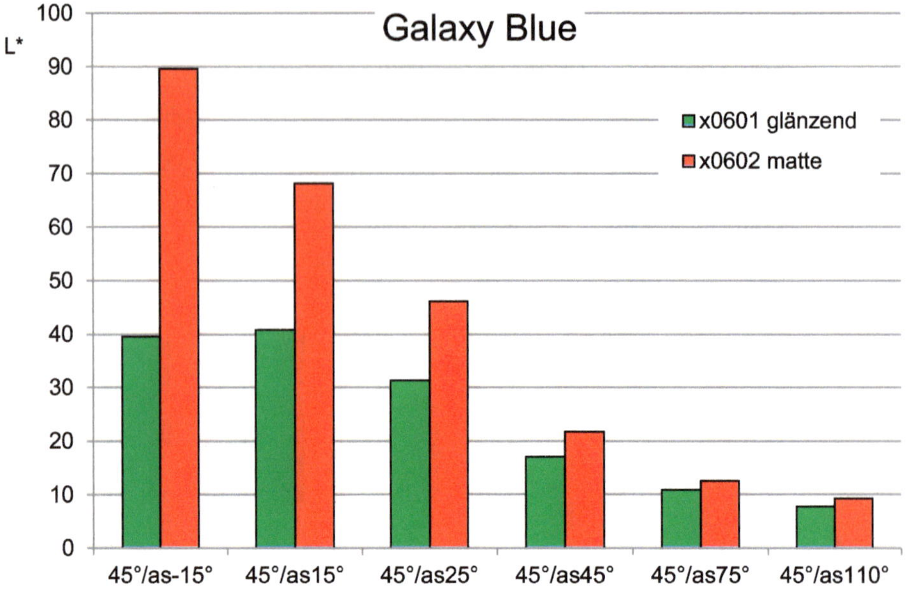

Figure 218: Brightnesses L* Galaxy Blue.

8 Transparency

Transparency is a particular issue when applying industrial and automotive coatings with effect pigments. On the one hand, many interference pigments – especially those based on mica, alumina, etc. – are transparent, but on the other hand, transparency is also a result of the base coats used. Since no base coat consists of 100 % pigments, there is always a greater or lesser degree of transparency in the base coat. This statement can also be confirmed by measurements of the same effect coating systems on black and white substrates.

Black and white transparency

Compared to a fine aluminum pigment, the brightness L* shows differences between the same base coat on a white and black substrate. In this example, the greatest differences are particularly evident in the medium gloss range (Figure 219). If these differences in brightness are compared with those of a coarse aluminum pigment, the differences between a white and black substrate become greater at angles as low as 25° (Figure 220).

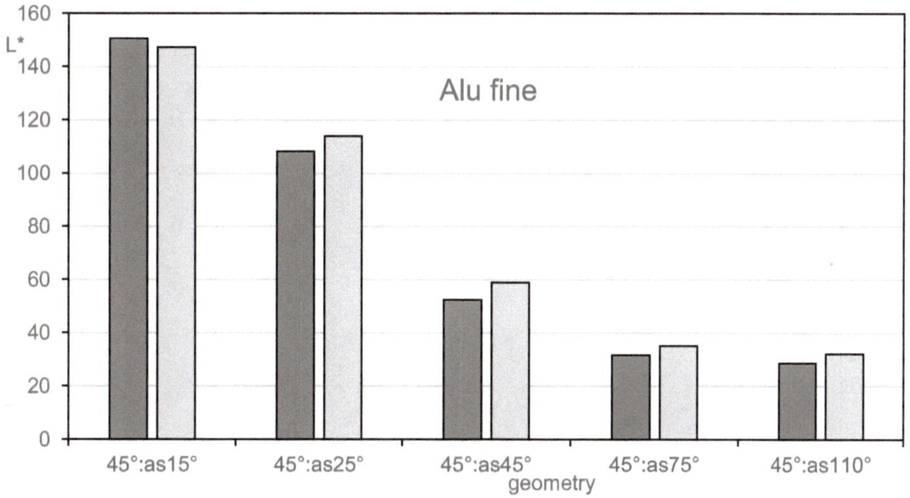

Figure 219: Brightnesses of fine aluminum on black and on white.

With transparent interference pigments, the differences between black and white substrates are, as expected, significantly greater. A white interference pigment of the fine fraction is darker overall than a comparable normal fraction. With the fine flake size, the differences between white and black substrates are greater at a distance from

https://doi.org/10.1515/9783111439204-008

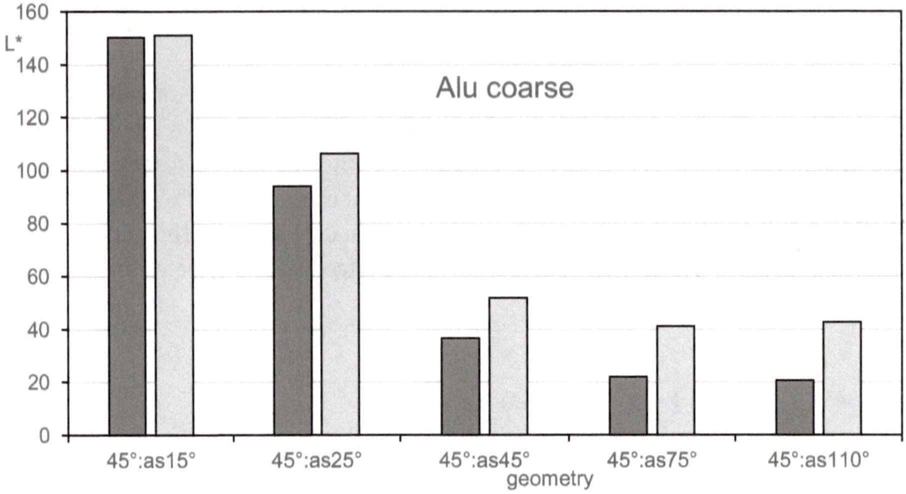

Figure 220: Brightnesses of coarse aluminum on black and on white.

specular than with the normal flake size. A fine blue interference pigment also shows that the brightness decreases more sharply at a distance from the gloss on a black substrate than on a white substrate (Figure 221).

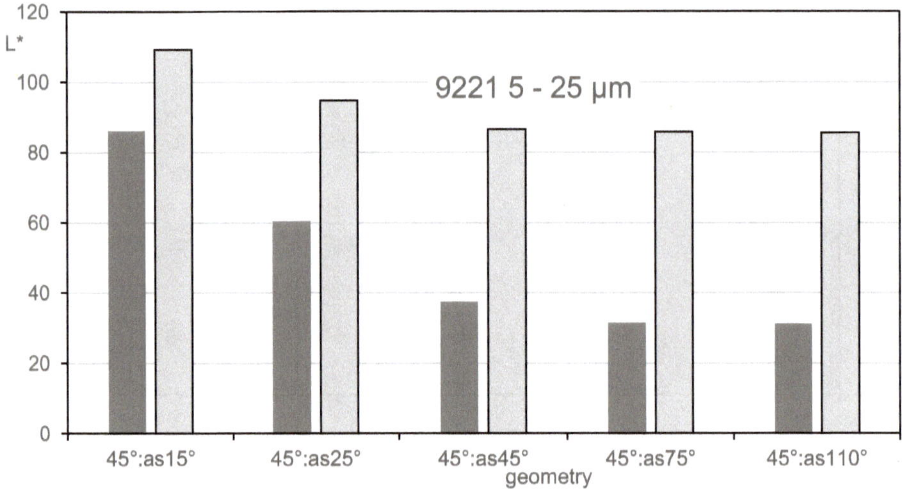

Figure 221: Brightnesses of fine Pearlblue 9221 on black and on white.

This also applies to the normal flake size of the blue interference pigment. Here, too, the brightness values decrease more sharply on a black substrate when viewed further away from the angle of gloss. The same applies to the example of the green interference

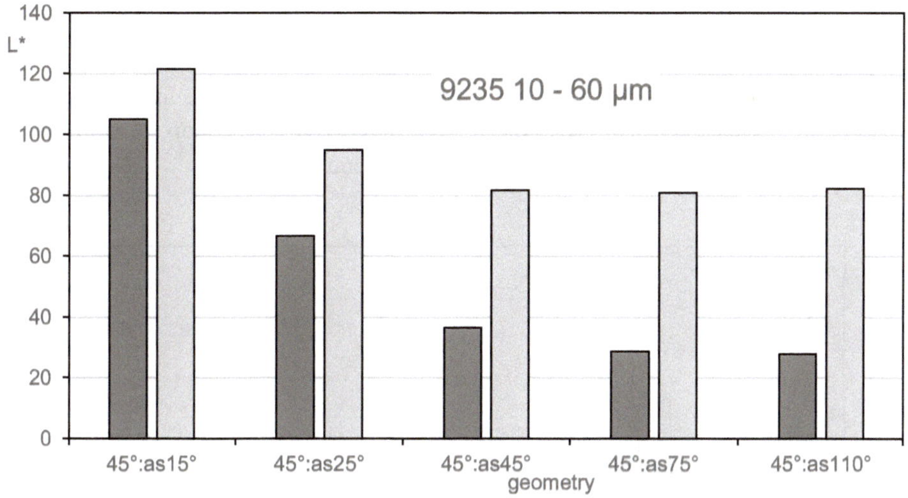

Figure 222: Brightnesses of Pearlgreen 9235 on black and on white.

pigment based on mica with titanium dioxide. Here, too, greater differences can be seen away from the gloss (Figure 222).

There are also differences between applying mica pigments with iron oxide or combinations of iron oxide and titanium dioxide coatings on black and white substrates. However, these differences are less significant than with transparent titanium dioxide micas due to their absorbent properties. Even with these types of mica – iron oxide/

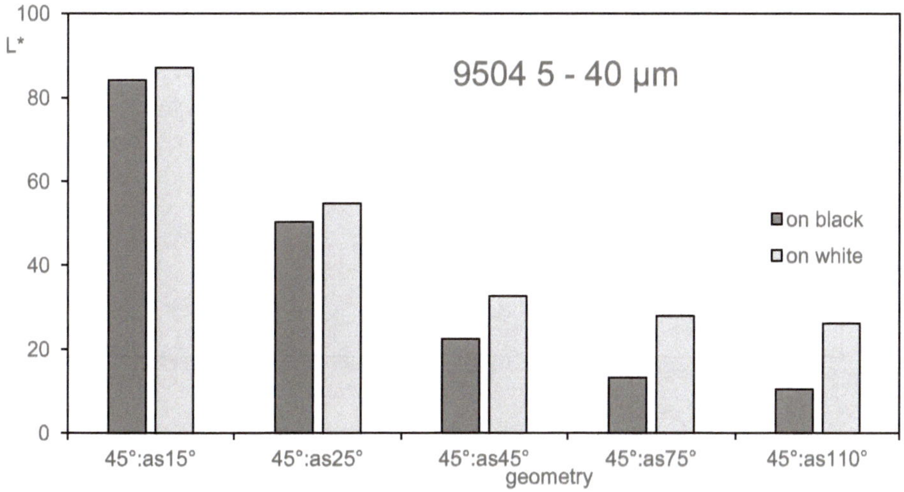

Figure 223: Brightnesses of 9504 on black and white.

titanium dioxide or iron oxide coated – the differences are less pronounced in the finer flake size than in the normal flake size (Figure 223).

This correlation with particle size is also evident when comparing Crystal Silver T60-10 and Microsilver T61-10. With the comparatively coarser version, the differences become greater the further away from specular you look.

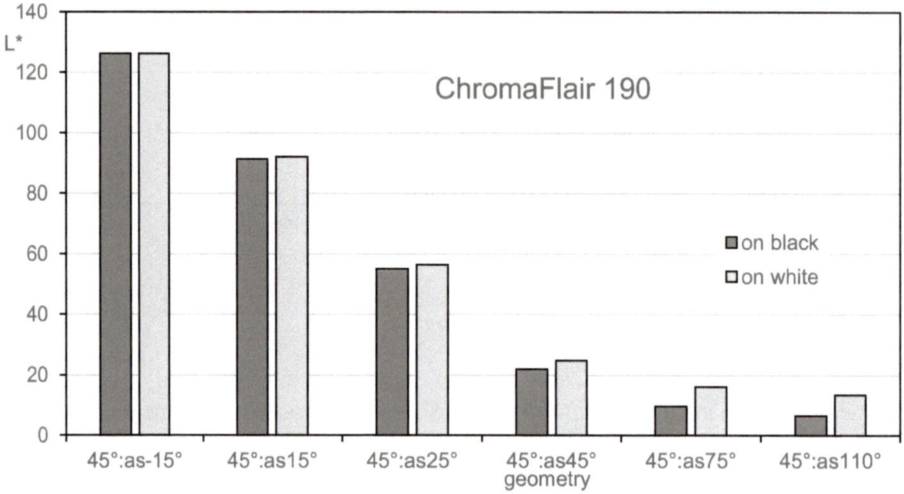

Figure 224: Brightnesses of Chromaflair 190 on black and on white.

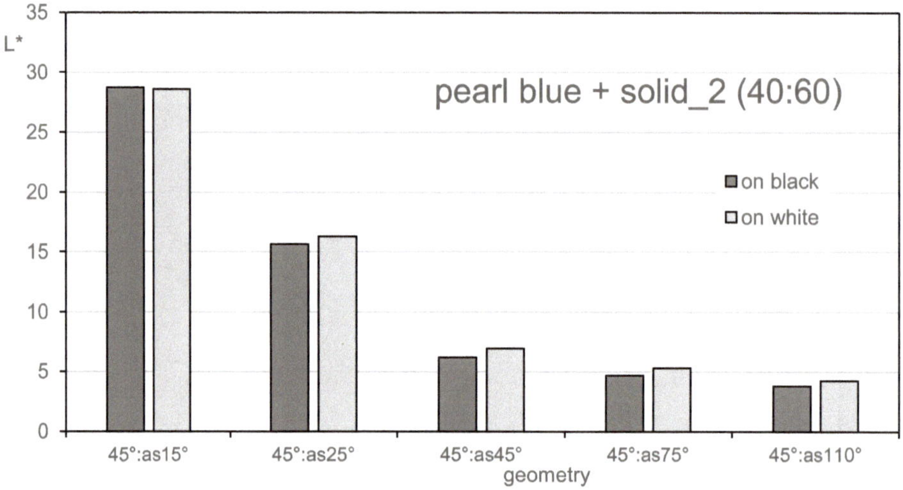

Figure 225: Brightnesses Pearlblue + solid_2 on black and on white.

Only when viewed further away from specular can differences in brightness between the white and black substrates be seen with ChromaFlair 190 (Figure 224). This interference pigment consists of an aluminum core.

When comparing mixtures of interference and aluminum pigments with colored pigments, the slight differences in the aluminum mixture at all measurement angles are noticeable in these examples, while the interference pigment blends show differences between the white and black substrates starting at 25° from specular (Figure 225).

Specialty

Some vehicles display a particularly “deep” and striking effect. In this case, a midcoat has been applied between the base coat and the clear coat. Alternatively, the base coat has been colored. In both cases, the observer perceives an intense color (Figure 226).

Figure 226: Brilliant red paintwork.

The influence of colored clear coat can be seen in the example of Ford’s Ruby Red. In two samples, the base coat is applied first, followed by the uncolored clear coat and then the colored clear coat (Figure 227). The differences are particularly evident in the measured chromaticity values C*. In Figure 228, the chroma C* is shown depending on whether it is observed at the window or in a light booth. The panels are tilted downwards and upwards. The angle between the light source and the observer always remains at 30°. The chroma C* increases significantly with the colored clear coat that has been additionally applied. Basically, the chroma C* increases the steeper the flanks of a reflection maximum are.

Figure 227: Red test coatings.

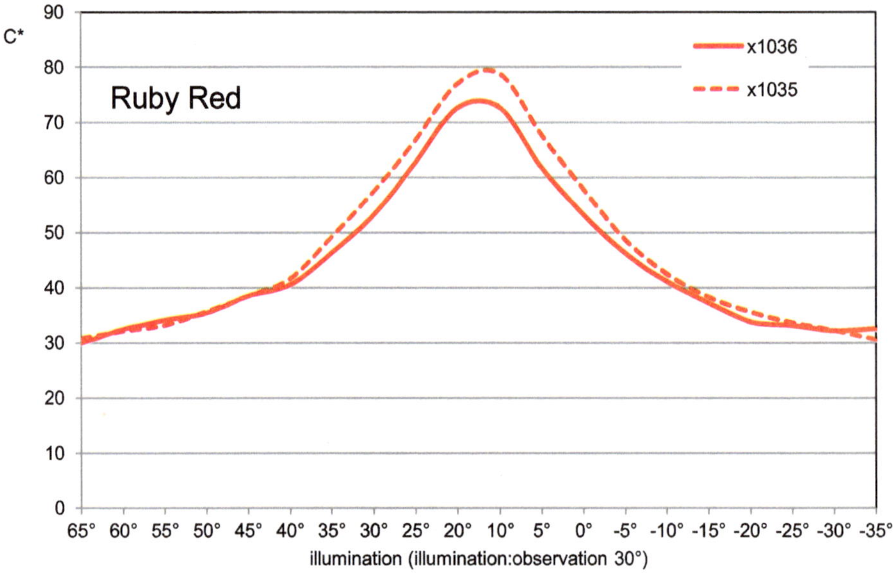

Figure 228: Observation at the window: Chroma C*.

Accordingly, the brightness levels in Figure 229 compiled. Here, the brightness is reduced by the colored clear coat.

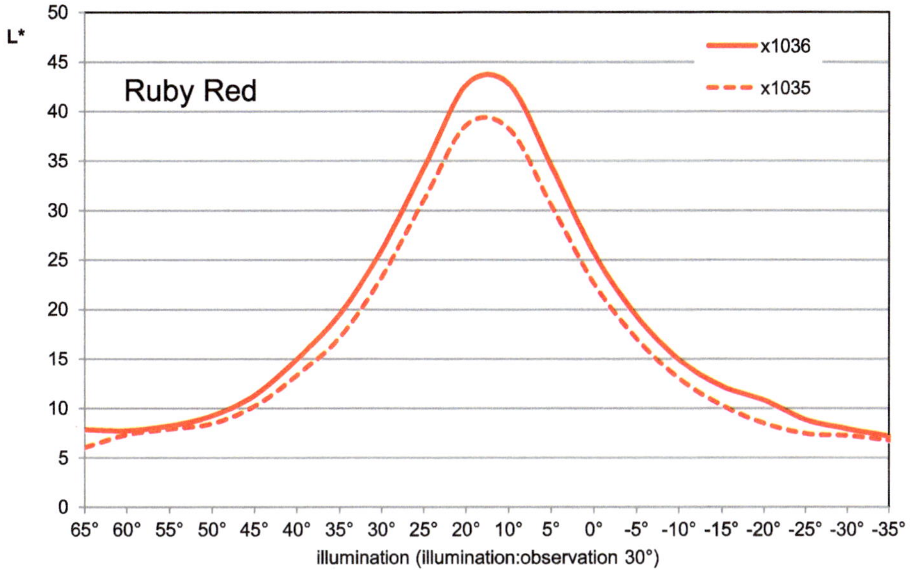

Figure 229: Observation at the window: Brightness L*.

9 From L*a*b* to RGB and back

The display of colors on a smartphone screen or a PC or laptop monitor is based on an RGB model. This consists of a cube with three edges representing the colors red, green, and blue. Their length ranges from 0 to 255 (8 bits). The specified RGB values represent exact colors, for example, 255 0 0 (red green blue) corresponds to red, 125 0 125 to dark violet, and 0 125 255 to greenish blue. White is 255 255 255 (red green blue) and black is 0 0 0. Yellow is mixed from red and green to 255 255 0 (Figure 230).

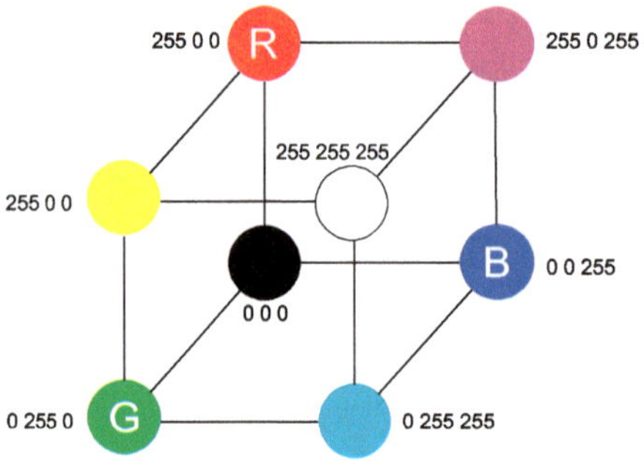

Figure 230: RGB gamut.

Microsoft and HP have defined the sRGB color space and Adobe has defined the AdobeRGB color space, whereby the Adobe RGB color gamut is slightly larger. The conversion starts with the standard color values X, Y, and Z according to standard formulas (Figure 231). If the results of the conversion for sRGB and Adobe RGB are entered into the chromaticity diagram, the limits of these color spaces become clear (Figure 232).

Interference and aluminum pigments push the limits of the RGB color space. In particular, color values close to gloss are outside the RGB space. Some software programs also "refuse" to accept higher L*a*b* values and convert them to RGB values. The limits of the RGB space can be demonstrated by experiments in Excel, for example. If you enter consecutive values for L*a*b* in the first columns and calculate AdobeRGB and sRGB in the following columns, you can see the values above 255 in the cells marked in red (Figure 233).

Unlike the RGB gamut, whose values are limited to between 0 and 255, the L*a*b* space is unlimited. This means that L* values for effect pigments, for example, can exceed 100.

For interference and aluminum pigments and corresponding paints and plastics, there are various solutions for "squeezing" them into the RGB color space. In contrast to colored pigments, they are described with multiple geometries ranging from near gloss at 15° from specular at –45° to far gloss at 110° aspecular (illumination angle 45°).

https://doi.org/10.1515/9783111439204-009

XYZ → Standard-RGB

```
//X, Y and Z input refer to a D65/2° standard illuminant.
//sR, sG and sB (standard RGB) output range = 0 ÷ 255

var_X = X / 100
var_Y = Y / 100
var_Z = Z / 100

var_R = var_X *  3.2406 + var_Y * -1.5372 + var_Z * -0.4986
var_G = var_X * -0.9689 + var_Y *  1.8758 + var_Z *  0.0415
var_B = var_X *  0.0557 + var_Y * -0.2040 + var_Z *  1.0570

if ( var_R > 0.0031308 ) var_R = 1.055 * ( var_R ^ ( 1 / 2.4 ) ) - 0.055
else                     var_R = 12.92 * var_R
if ( var_G > 0.0031308 ) var_G = 1.055 * ( var_G ^ ( 1 / 2.4 ) ) - 0.055
else                     var_G = 12.92 * var_G
if ( var_B > 0.0031308 ) var_B = 1.055 * ( var_B ^ ( 1 / 2.4 ) ) - 0.055
else                     var_B = 12.92 * var_B

sR = var_R * 255
sG = var_G * 255
sB = var_B * 255
```

Figure 231: Calculating RGB from XYZ.

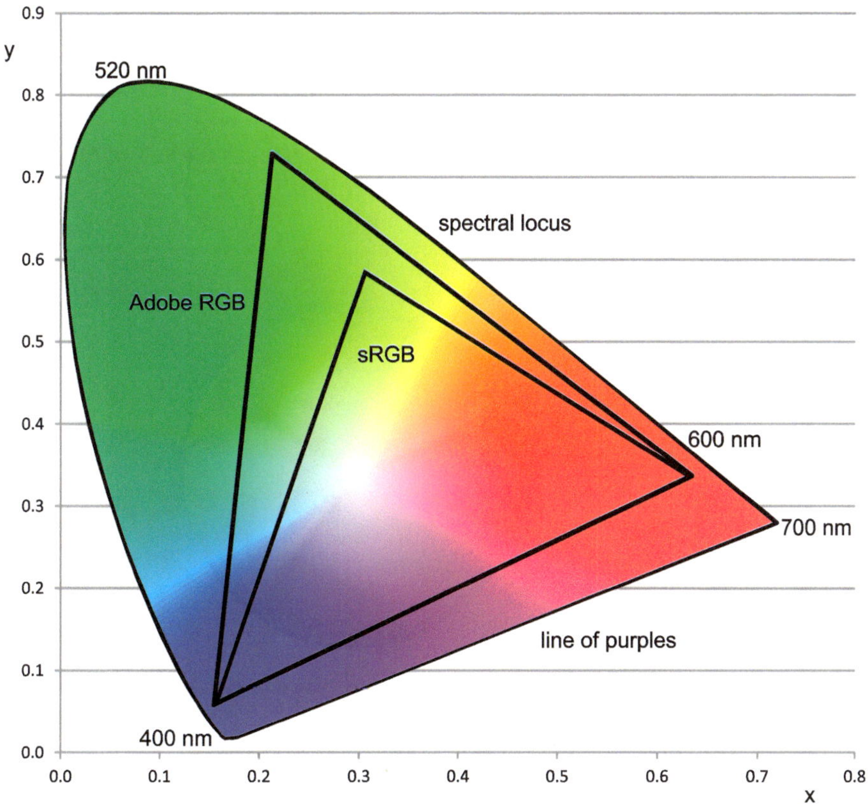

Figure 232: The chromaticity diagram with the sRGB and AdobeRGB color gammuts.

			Adobe RGB			sRGB		
L*	a*	b*	R	G	B	R	G	B
10	-40	-40	0	47	82	0	43	82
20	-40	-40	0	66	105	0	64	107
30	-40	-40	0	88	129	0	88	132
40	-40	-40	0	112	155	0	113	158
50	-40	-40	0	138	182	0	139	185
60	-40	-40	0	164	210	0	166	213
70	-40	-40	43	192	239	0	193	241
80	-40	-40	84	221	269	0	222	270
90	-40	-40	116	250	300	0	251	300
100	-40	-40	148	281	331	0	280	330
110	-40	-40	179	312	364	37	310	360
120	-40	-40	210	345	397	105	341	392
130	-40	-40	242	378	431	149	372	423
140	-40	-40	274	412	466	188	403	455
150	-40	-40	307	446	501	224	435	488
160	-40	-40	341	482	537	259	468	521
170	-40	-40	375	518	574	294	501	554
180	-40	-40	410	554	612	328	534	588

Figure 233: Unrepresentable L*a*b* values (red).

Both the calculation of CIELAB values and that of OKLAB and RGB values starts with the color values XYZ. These are determined using the reflection values, spectral power distribution, and the color-matching functions (Figure 234).

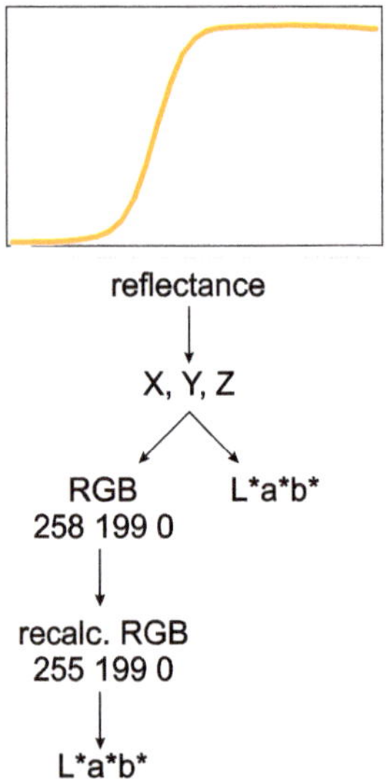

Figure 234: RGB and L*a*b* are based on XYZ.

Example 1: All RGB values above 255

If an RGB value is greater than 255, it is reduced to 255: RGB 397 200 268 becomes RGB 255 200 255, or RGB 222 283 205 becomes RGB 222 255 205 (Figure 235). This "reduction" occurs automatically in some programs. Other programs "refuse" to accept RGB values higher than 255.

				calculated from L*a*b* by measurement		
Stellar Green	L*	a*	b*	R	G	B
45°:as-15°	132.40	-63.18	9.61	265	394	334
45°:as15°	110.73	-51.17	17.75	226	316	251
45°:as25°	67.96	-22.23	11.59	144	175	143
45°:as45°	40.19	-4.03	-5.88	87	97	103
45°:as75°	35.37	-1.32	-10.98	76	85	99
45°:as110°	36.19	-2.17	-11.42	77	88	102
15°:as15°	102.08	-41.69	28.62	222	283	205

all values above 255 reduced to 255				example1			
R	G	B	L*	a*	b*	C*	h°
255	255	255	100	0	0	0.10	0.00
226	255	251	97.2	-13.8	-2.4	14.01	189.87
144	175	143	68	-22.5	12.4	25.69	151.14
87	97	103	40.1	-4.2	-5.6	7.00	233.13
76	85	99	35.2	-1.4	-10.6	10.69	262.48
77	88	102	36.3	-2.5	-10.9	11.18	257.08
222	255	205	95.7	-24.2	19.7	31.20	140.85

all values above 255 reduced to 255 by factor					example2			
Faktor	R	G	B	L*	a*	b*	C*	h°
1.55	172	255	216	91.9	-46.2	7.8	46.85	170.42
1.24	182	255	203	92.3	-44.3	15.5	46.93	160.72
1.00	144	175	143	68.1	-22.5	12.4	25.69	151.14
	87	97	103	40.1	-4.2	-5.6	7.00	233.13
	76	85	99	35.2	-1.4	-10.6	10.69	262.48
	77	88	102	36.3	-2.5	-10.9	11.18	257.08
1.11	200	255	185	93.4	-38.7	26.9	47.13	145.20

all values calculated by max. factor					example3			
Faktor	R	G	B	L*	a*	b*	C*	h°
1.55	172	255	216	91.9	-46.2	7.8	46.85	170.42
	146	205	162	76.3	-38.2	13.8	40.62	160.14
	93	113	93	45	-16.1	8.6	18.25	151.89
	56	63	67	24.8	-3.3	-4.3	5.42	232.50
	49	55	64	21.2	-1.2	-7.7	7.79	261.14
	50	57	66	22.1	-1.8	-7.9	8.10	257.16
	144	183	133	69.8	-30.1	21	36.70	145.10

Figure 235: Various calculations.

The disadvantage of this approach is that it "distorts" the color.

Example 2: Factor for a geometry

Here, the highest value of an RGB color greater than 255 is determined. This value is divided by 255 and the other two RGB values are divided by this determined factor. From RGB 397 200 268, the factor 1.56 is calculated, which then results in the RGB values 255 128 172. Or from RGB 222 283 205, the calculated factor 1.11 results in the RGB values 200 255 185.

This method has the disadvantage that the values calculated in this way for the geometries close to gloss are too close to the following geometries in the RGB space, and the color gradient no longer appears plausible.

Example 3: Factor for all geometries

Here, the highest value of all geometries is determined, and this value is divided by 255. This factor is then applied to all RGB values. In this way, the result – the color gradient across the geometries – appears more plausible.

The RGB values for the 45°:as–15° geometry are reduced by a factor of 1.55 from RGB 265 394 334 to RGB values 172 255 216, and the RGB values for the 45°:as15° geometry are reduced from 226 316 251 to 146 205 162.

This creates a plausible color difference between the geometries because the factor is applied to all RGB values.

Factor relating to brightness L*

With this method, the determined factor is calculated specifically for all geometries with the respective brightness L*. This makes the resulting RGB values brighter, i. e., their individual values become larger.

Aluminum

The representation of the brightness values from the measurement and the calculated examples shows the decrease in brightness with geometries that are further away from the gloss angle (gray columns). In examples 1 and 2, the L* values are reduced to 100 near the gloss angles. From geometry 45°:as45° onwards, the calculated L* values are

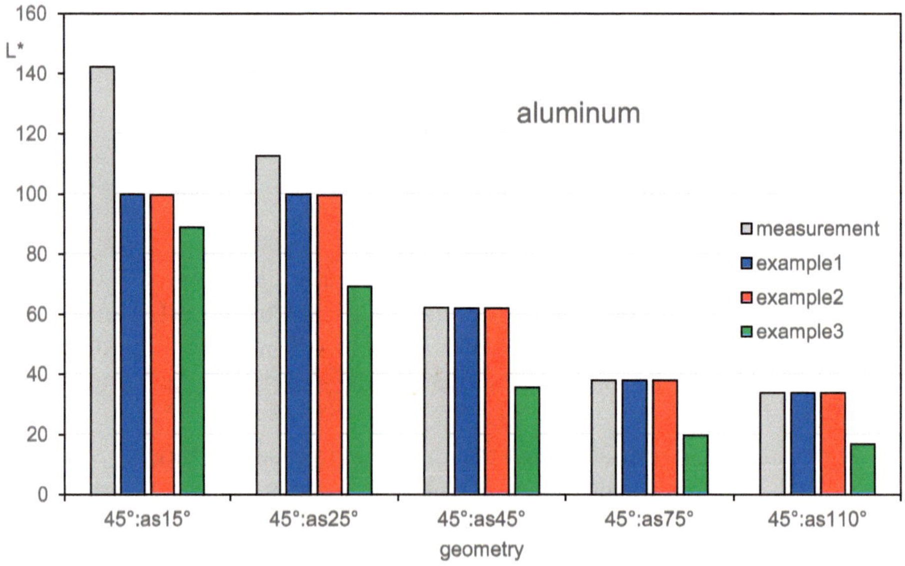

Figure 236: Recalculated brightness values L*.

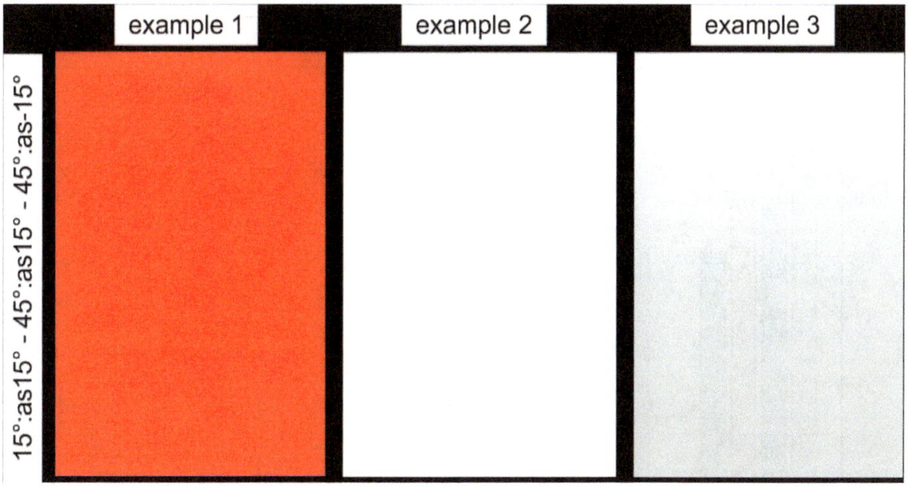

Figure 237: Representation of the original values of aluminum outside the RGB space (red) and recalculated values.

like the measured values. In example 3, the brightness values L* are shifted parallel to the measured values (Figure 236).

Figure 237 displays the RGB values as a color gradient from geometry 15°:as15° to 45°:as15° to 45°:as–15°. Example 1 is colored red; here, all RGB values are reduced to 255.

Pearlwhite

The measured and calculated brightness values L* for pearl white are similar to those for aluminum (Figure 238). When displaying the RGB values as a color gradient, example 3 appears darker (Figure 239).

Stellar Green

Stellar Green is a green interference pigment whose color shifts from yellow-green under steep illumination to blue-green under flat illumination. The conversions and reductions change the color values, which differ significantly from the original color (Figure 240).

The converted brightness values L* differ at 45°:as15° but then match the measured values for examples 1 and 2. The values for example 3 decrease in parallel with the measured L* values (Figure 241).

Clear differences can be seen in the chroma C*. Here, the values in example 1 at 45°:as15° are striking (Figure 242).

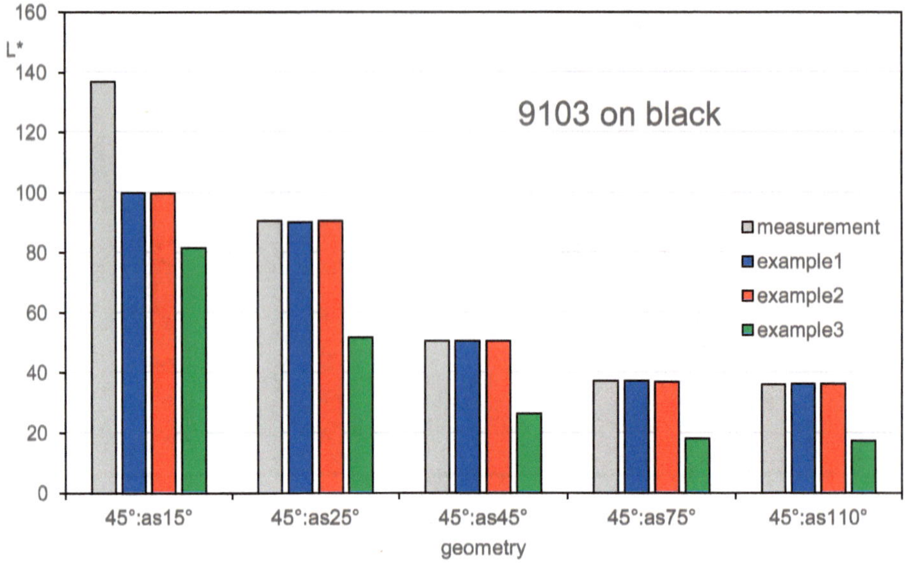

Figure 238: Measured and recalculated L* values for Pearlwhite 9103.

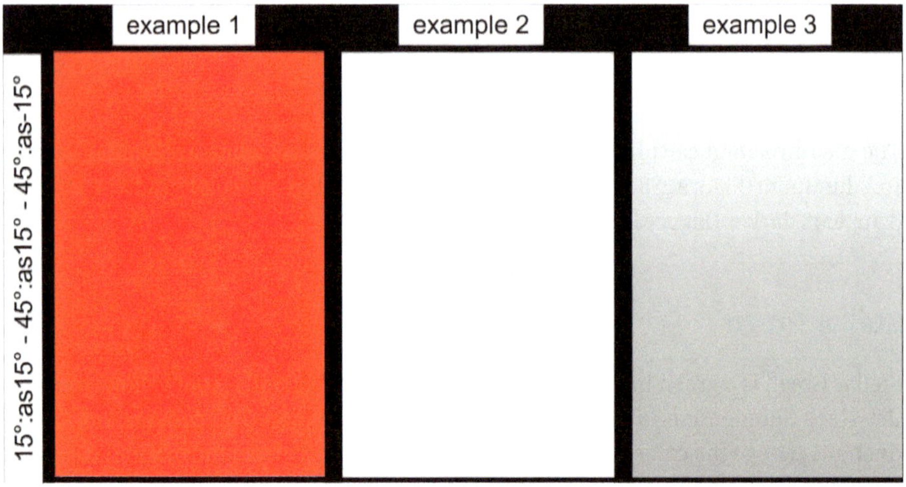

Figure 239: Representation of the original values of 9103 outside the RGB space (red) and recalculated values.

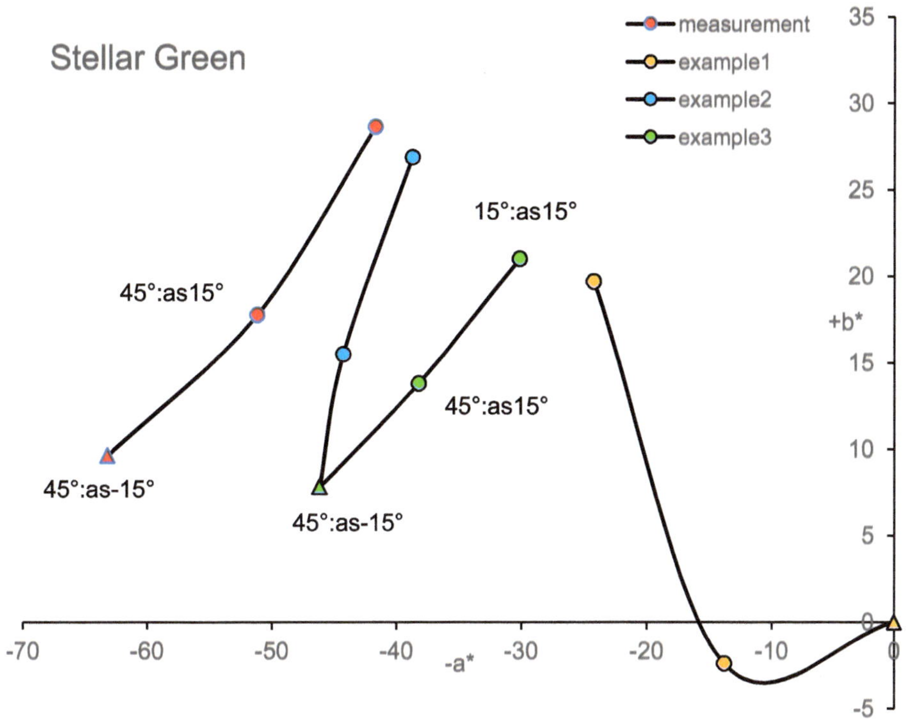

Figure 240: Measured and recalculated color values for Stellar Green.

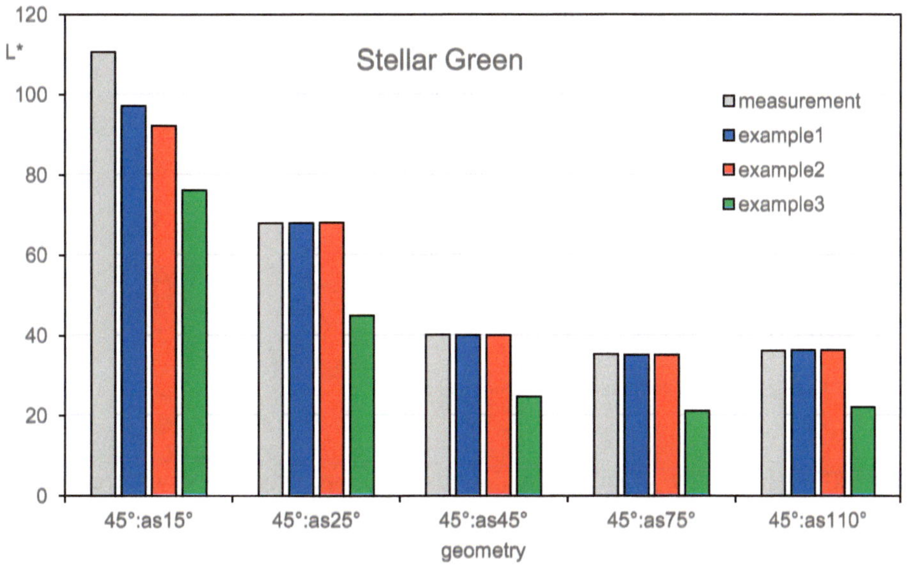

Figure 241: Measured and recalculated L* values for Stellar Green.

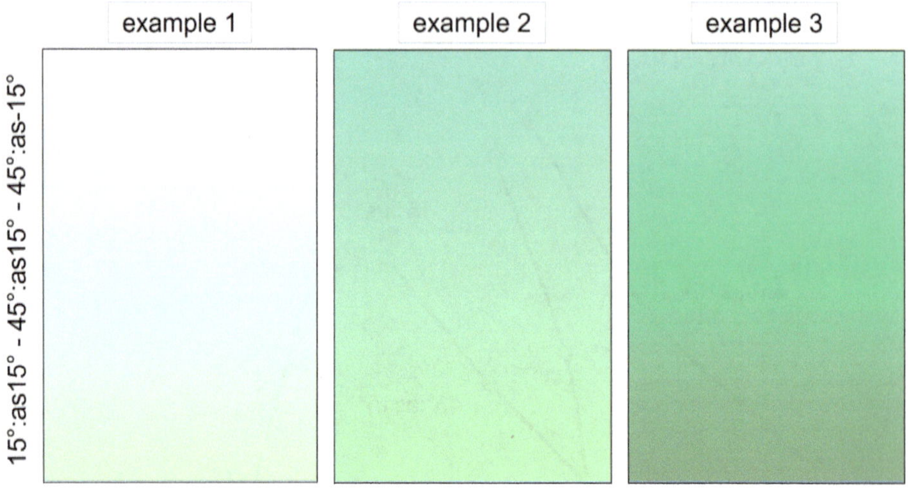

Figure 242: Presentation of the original values from Stellar Green and recalculated values.

Aluminum + blue

This series of blends between an aluminum pigment and solid_2 (blue) was recorded colorimetrically (Figure 243).

panel				
91014			aluminum	100
91013	solid_2	2	aluminum	98
91012	solid_2	5	aluminum	95
91011	solid_2	10	aluminum	90
91010	solid_2	20	aluminum	80
91009	solid_2	30	aluminum	70
91008	solid_2	40	aluminum	60
91007	solid_2	50	aluminum	50
91006	solid_2	60	aluminum	40
91005	solid_2	70	aluminum	30
91004	solid_2	80	aluminum	20
91003	solid_2	90	aluminum	10
91001	solid_2	100	aluminum	0

Figure 243: Percentages of blend series aluminum + blue.

The RGB values were calculated and adjusted accordingly. These were converted back into sRGB and Adobe RGB values. The brightness diagram shows the continuous decrease in the measured values. The converted values do not reflect this decrease (Figure 244). The same can be seen in the chroma diagram, where the recalculated C* values only appear when there is a larger proportion of color pigment blends (Figure 245).

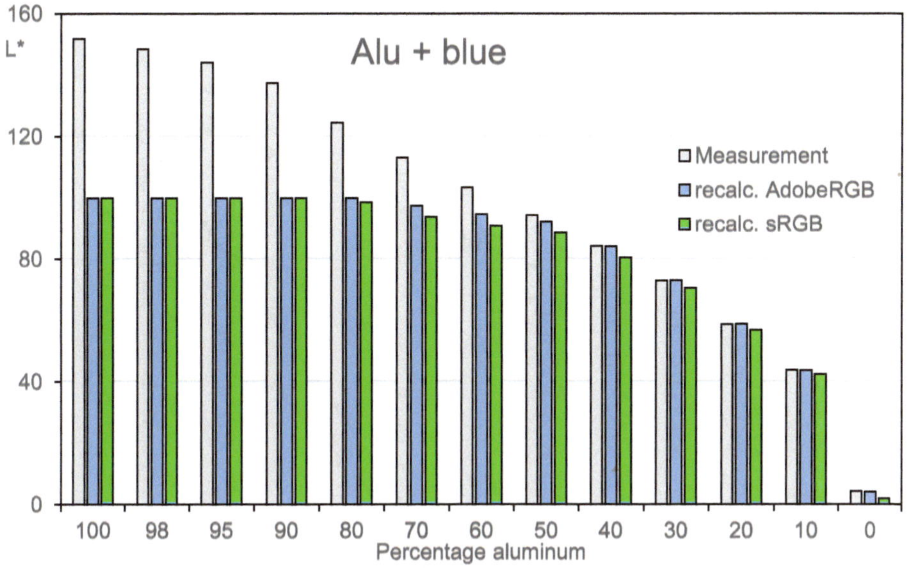

Figure 244: Measured values and recalculated values L*.

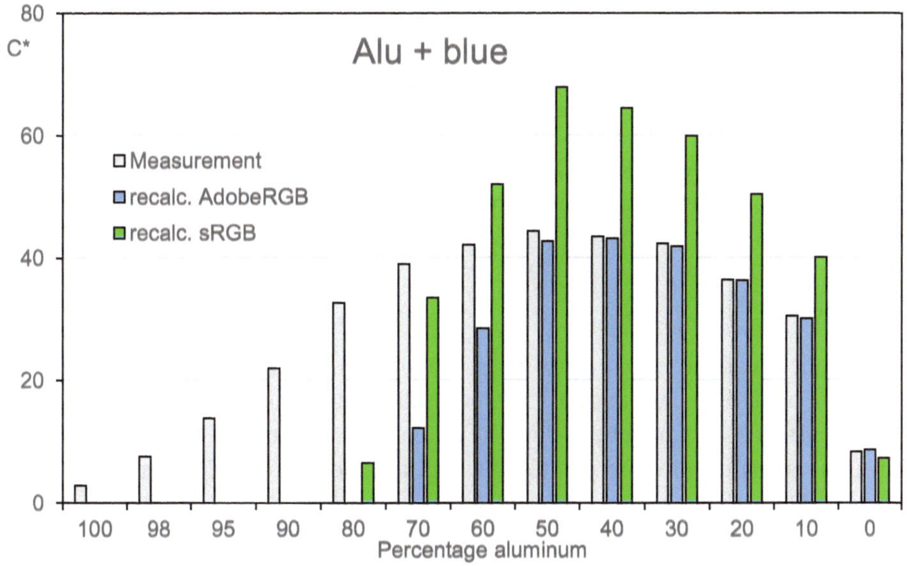

Figure 245: Measured values and recalculated values C*.

The conversions are much more dramatic when displaying the a*b* color values (Figure 246). The measured values typically range from aluminum to blue (solid_2) via the turning point. The newly calculated color values a*b* are based on the values con-

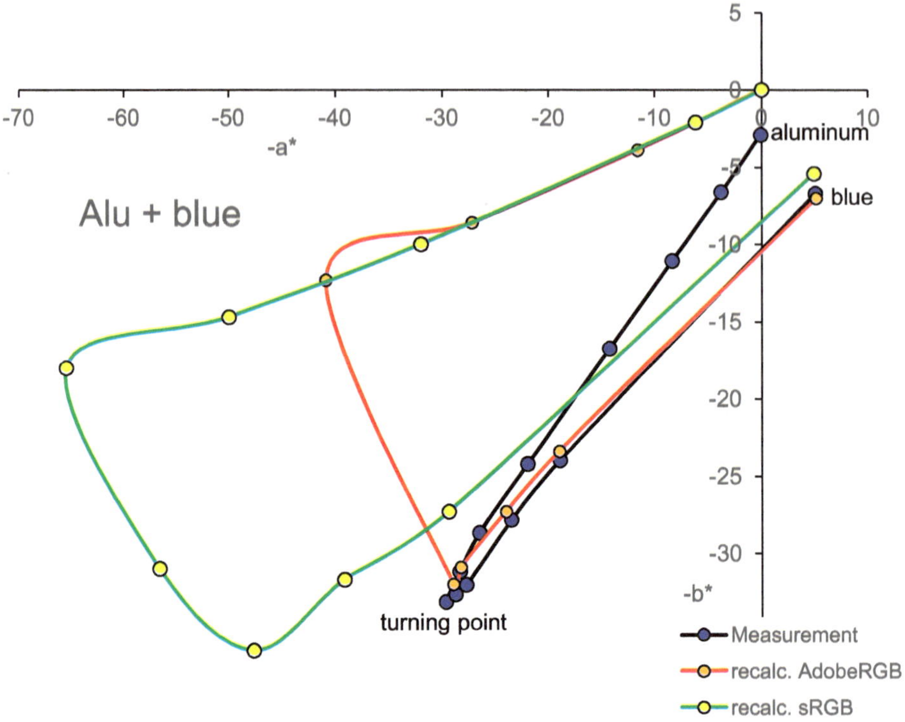

Figure 246: Measured values and recalculated color values.

verted in sRGB and AdobeRGB. There are significant differences between these and the measured color values, which lead to incorrect representations.

ChromaFlair 190

This pigment has the widest color range, from green under steep illumination to orange, red, violet, and blue, and back to green under flat illumination, with 15° aspecular. If you were to lay a painted panel on the ground in Hammerfest, Norway, on June 21, when the sun is at its highest point, you would see colors ranging from yellow to violet (Figure 247). At the same time, blue would also be visible in Hamburg (Germany) and green in Key West (Florida). This is made possible by the different positions of the sun in these locations at the same time (Figure 248).

In the a*b* diagram, (Figure 249) shows this color gradient from steep illumination (20°:as15°) to flat (65°:as15°). From 45°:as15° onwards, the aspecular line is entered, which reflects the reproduction of the portable instruments.

Figure 247: Same time, different places, different colors.

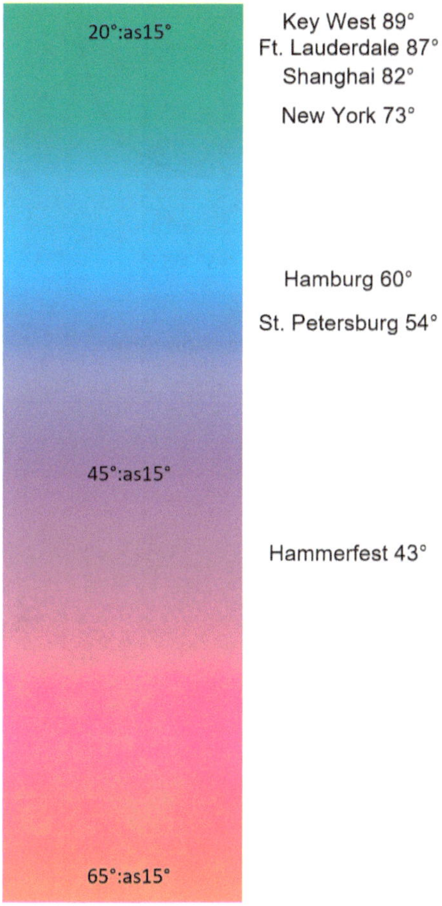

Figure 248: Colors of the Chromaflair 190 depending on location and sun position.

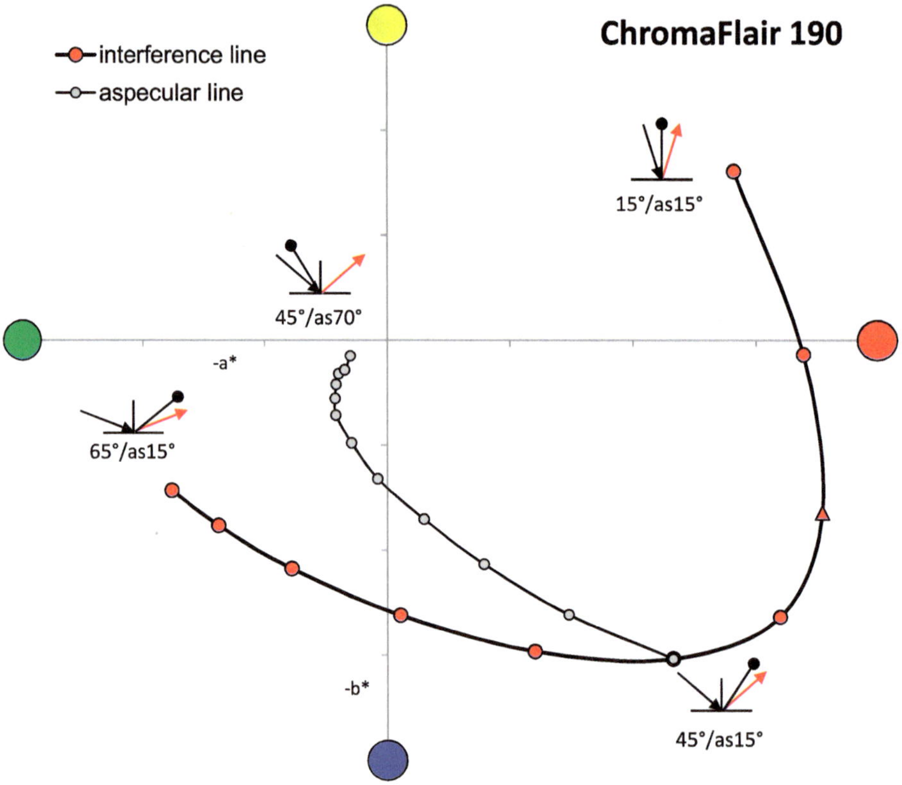

Figure 249: The interference line of Chromaflair 190 runs through almost the entire color space.

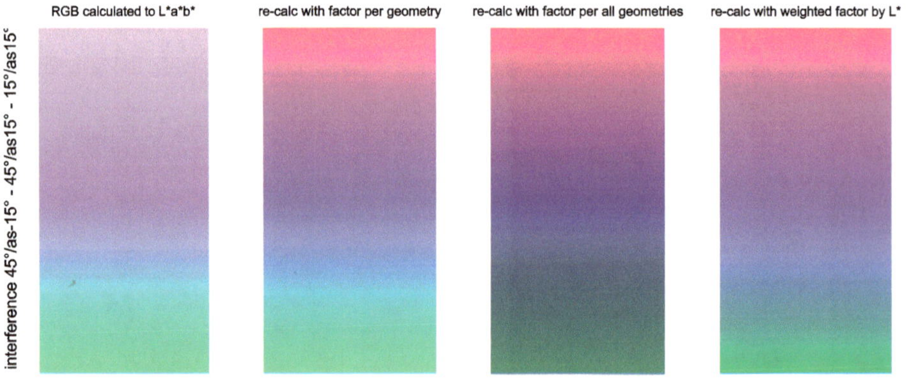

Figure 250: Calculated interference color gradients.

Figure 251: Calculated aspecular color gradients.

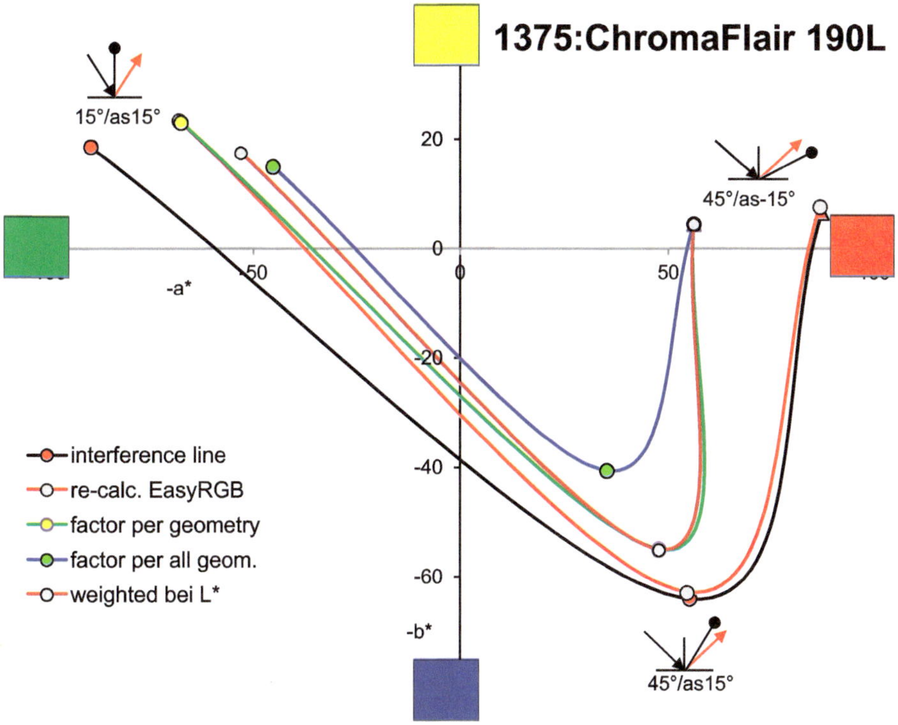

Figure 252: Calculated color values of Chromaflair 190.

Conversion according to the pattern described reveals different RGB color gradients for the interference line from 20°:as15° to 45°:as15° to 65°:as15° (Figure 250). The differences are also clear in the aspecular lines (Figure 251).

The effects of the various conversions are also visible in the a*b* diagram. If all measured values are converted using the factor determined from the maximum value, the calculated interference line is shifted parallel to the measured line (Figure 252). The a*b* values based on the weighted L* values are closer to the measured values in the middle geometry range 45°:as15°.

In principle, there is no reason not to convert L*a*b* or XYZ values based on reflection measurements into RGB values. However, the limits of the RGB color space should be considered. It is not only interference and aluminum pigments that have glossy color values outside the RGB color space. Some very intense and bright colored pigments can also lie outside the RGB color space.

This limitation should be considered when using the RGB color space on a laptop or smartphone screen.

10 Microscope

The analysis of an unknown paint or plastic with effect pigments begins with examination under a microscope. This allows the basic properties of an effect pigment to be identified. Aluminum pigments and, in some cases, interference pigments are "opaque," which means they cannot be viewed in transmission. The dark field setting is suitable here. Transparent interference pigments can be examined in transmission. Here, they reveal their structures in the transmission color.

For intensive analysis with a microscope, it is advisable to place samples on slides, for example, paints from a mixing system. If the microscope has a camera, an image database can also be created.

When examining interference pigments under a microscope, it is advisable to view them several times at different points, especially in the case of wet-chemical pigments. In wet-chemical manufacturing processes, the resulting color depends on the duration of the process, and thus on the layer thickness. Therefore, in addition to pigments with the desired main color, pigments with a lower or higher layer thickness may also be produced.

Interference pigments based on natural mica have irregular structures. The coating of natural mica with titanium dioxide, for example, is uniform and regular. Irregularities relate to the mica carrier flakes, which is why they do not show a uniform color under the microscope (Figures 253–272).

Reflected light—transmitted light

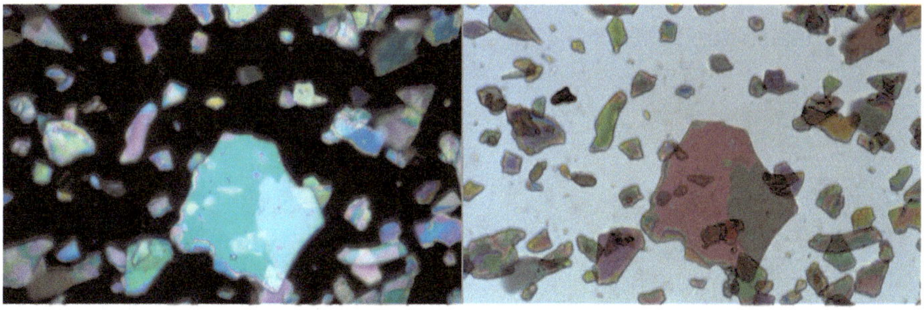

Figure 253: Stellar Green.

https://doi.org/10.1515/9783111439204-010

Figure 254: Colorstream Viola Fantasy.

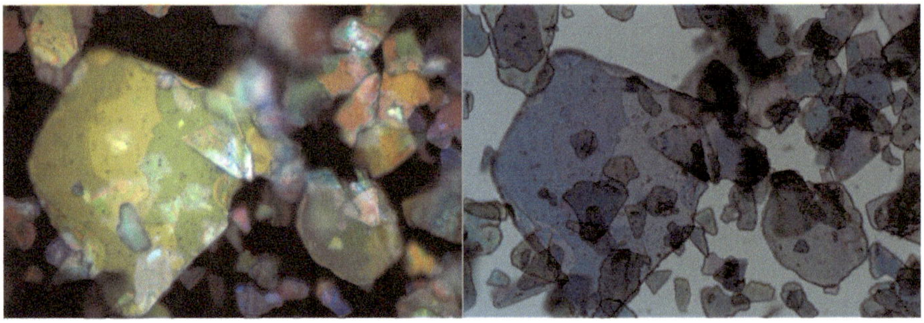

Figure 255: Sunbeam Gold.

Reflected light—darkfield light

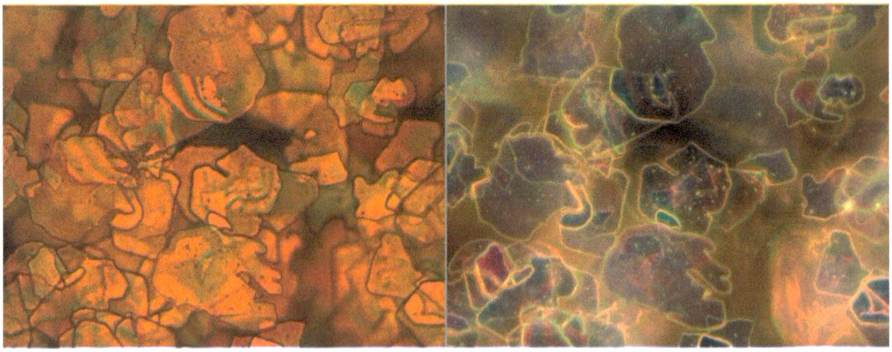

Figure 256: Sunbeam Gold T60-20.

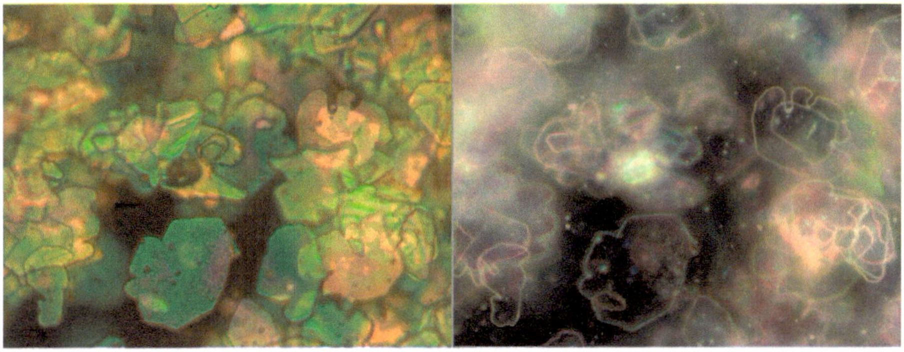

Figure 257: Stellar Green T60-24.

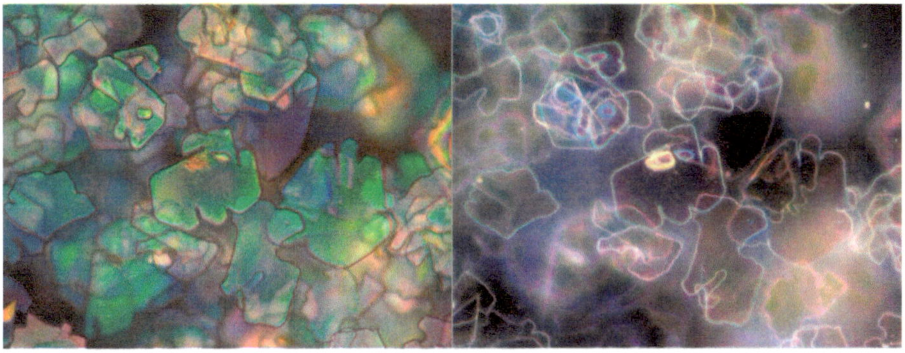

Figure 258: Galaxy Blue T60-23.

Figure 259: SpectraFlair Silver 1400.

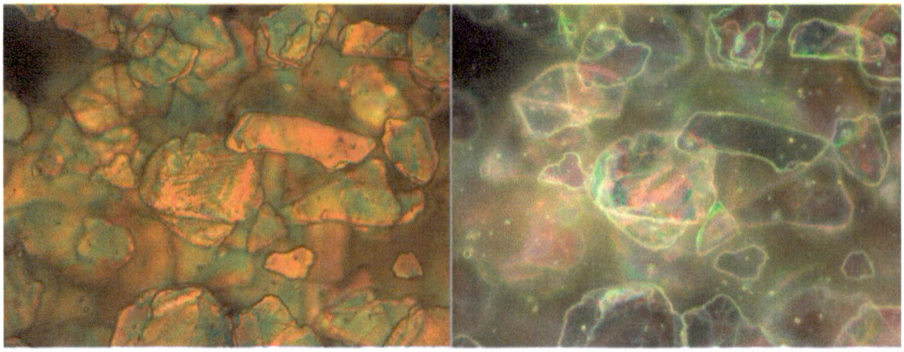

Figure 260: Pearlgreen 9235.

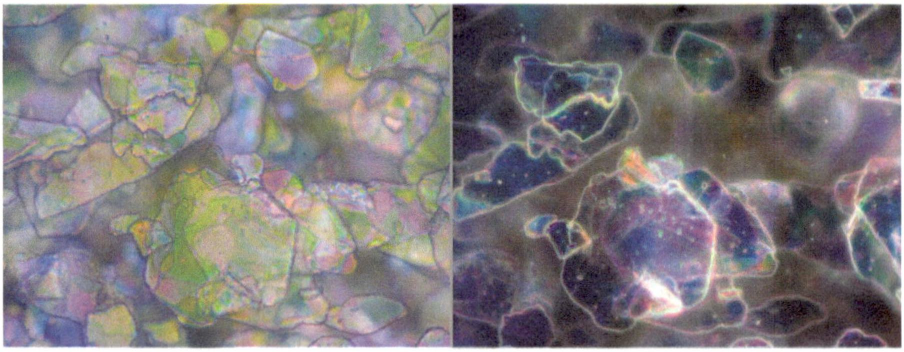

Figure 261: Pearl Gold 9205.

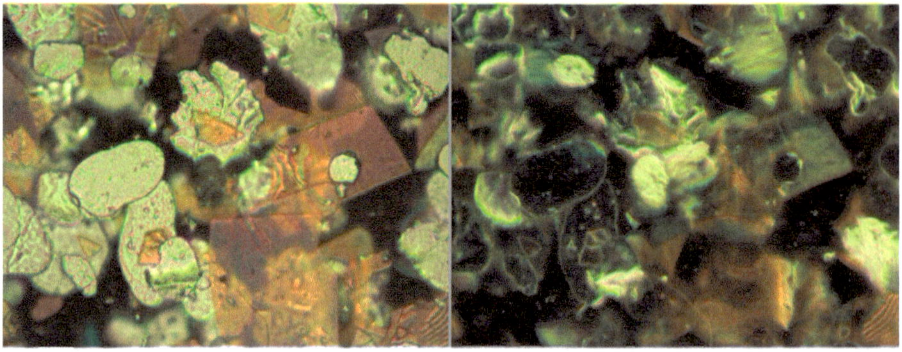

Figure 262: Alu + Autumn Mystery 60:40.

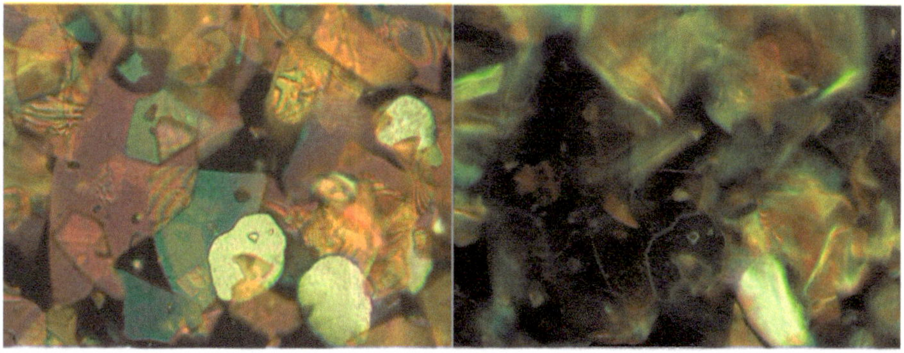

Figure 263: Alu + Autumn Mystery 10:90.

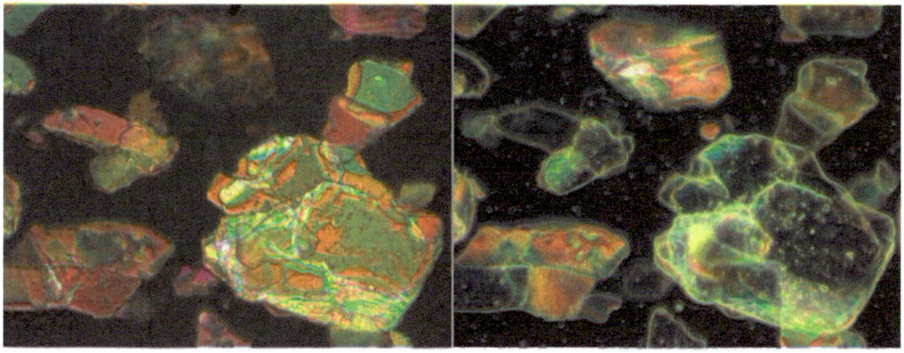

Figure 264: Pearl Lila 9217.

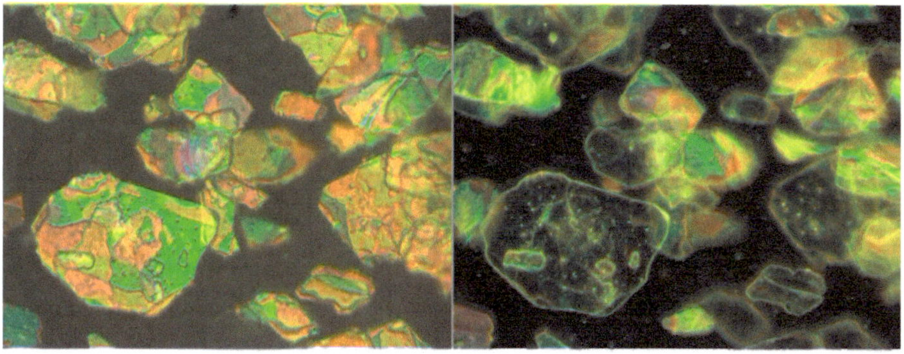

Figure 265: Gold Pearl 9300.

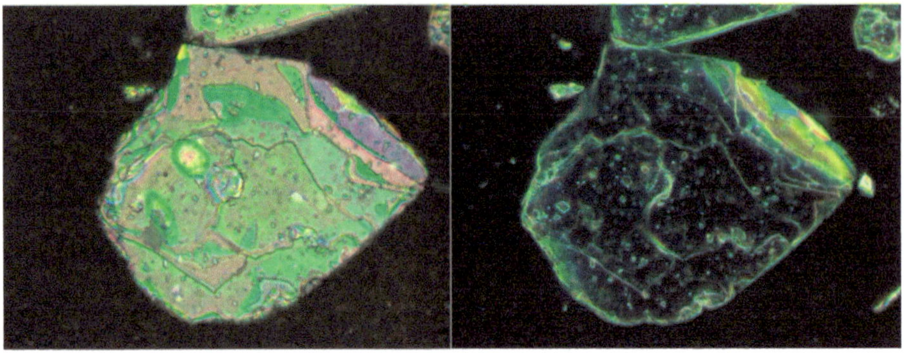

Figure 266: Flash Pearl 9153.

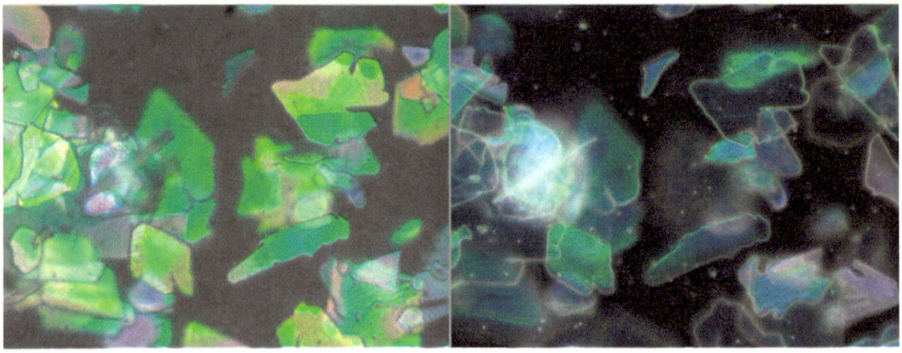

Figure 267: Stellar Green T60-24.

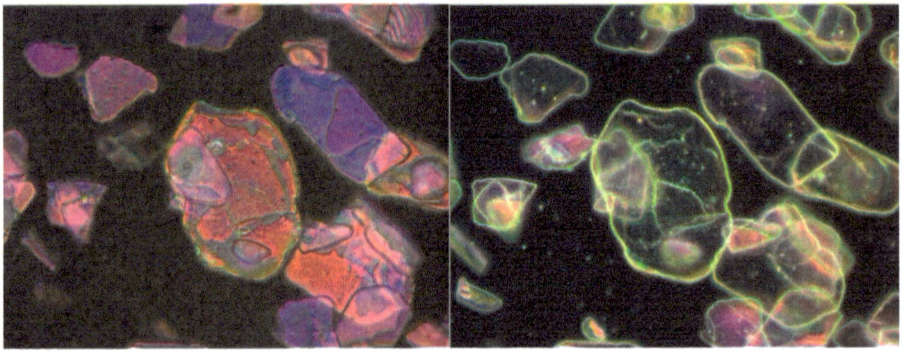

Figure 268: Pyrisma Color Space Magenta.

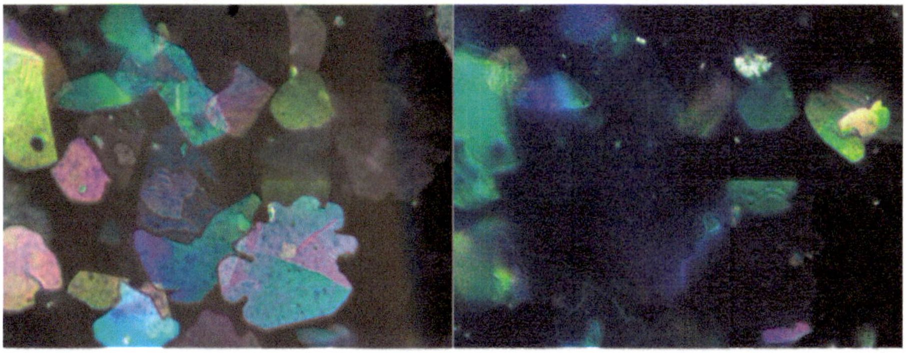

Figure 269: Northern Light Grey.

Figure 270: Crystal Silver T60-10 w/o effect layer.

Figure 271: Crystal Silver T60-10 with effect layer.

Figure 272: Colorstream Arctic Fire T20-02.

Index

https://doi.org/10.1515/9783111439204-011